Humanistic Management and Sustainable Tourism

Tourism is a fast-growing and changing industry, which has become a driver of economic development in both developed and underdeveloped countries. While the tourism industry's potential for shared value creation and sustainable development is acknowledged, the concerns around the environmental and social pressures remain a challenge for businesses, organizations, and destinations. This is because sustainable tourism arguably conflicts with the predominant neoliberal structure of the economy and with the hierarchical, profit- and consumption-driven societies. The emphasis on competition, growth, and profitability may undermine economic viability itself by consuming unreproducible resources and by undermining the six essential elements—dignity, people, prosperity, social justice, planet, and partnership—that are conceptually linked to sustainable development. The crises recurrently challenging the global travel and tourism environment, including climate change, bushfires, extreme weather disasters, pandemics, and the financial crisis, show the weaknesses of neoliberal approaches and the collective economic dependency of countries on tourism that is vulnerable, if not completely unsustainable. This vulnerability asks for understanding that the collective future depends on developing entirely new approaches and interpretations of tourism to effectively respond to human, societal, social, and climate challenges.

This book offers a novel and original perspective entailing the application of a humanistic management approach to sustainable tourism, which is centered on the value of human life, the protection of human dignity, and the promotion of well-being. Multiple theoretical approaches, methods, and practical cases, on an international scale, shed light on shared value creation and human dignity as a necessary condition for its achievement in different contexts. Implicitly and explicitly, they respond to the current urgency to implement strategies to recover from the worldwide impact of the pandemic crisis and to provide a vision of what tourism could and should be when it recovers. This book will be of interest to researchers, academics, professionals, and postgraduates in the fields of management, sustainability, and tourism development.

Maria Della Lucia is Associate Professor of Tourism and Business Management at the University of Trento, Italy. Her main areas of research, teaching, and training include local development and sustainability, destination management and governance, culture-led regeneration, creative cities and creative tourism, digital and social media marketing, and economic impact analysis as investment decision-making tools. She has published her research in prominent journals.

Ernestina Giudici is a retired Full Professor of Management and Business Communication at the University of Cagliari, Italy. Her main areas of research include business communication, storytelling and digital storytelling, humanistic management and integrity, corruption, ethical sustainable development, sustainability, and neurotourism. She is part of the managing team of the International Humanistic Management Association and has published widely in various prestigious journals.

Humanistic Management
Series Editors: Michael Pirson, Erica Steckler, David Wasieleski, Benito Teehankee, Ricardo Aguado and Ernestina Giudici

Humanistic Management draws together the concepts of social business, sustainability, social entrepreneurship, business ethics, conscious capitalism and cooperative capitalism to present a new humanistically-based research paradigm. This new paradigm challenges the prevailing neo-liberal 'economistic' approach that dominates twentieth-century management theory and practice, and instead emphasizes the need to protect human dignity and wellbeing as well as economic drivers.

Aesthetics, Organization, and Humanistic Management
Edited by Monika Kostera and Cezary Wozniak

Motivation in Organizations
Searching for a Meaningful Work-Life Balance
Manuel Guillén

Humanistic Management and Sustainable Tourism
Human, Social and Environmental Challenges
Edited by Maria Della Lucia and Ernestina Giudici

Humanistic Management and Sustainable Tourism

Human, Social and Environmental Challenges

Edited by Maria Della Lucia and Ernestina Giudici

Routledge
Taylor & Francis Group
NEW YORK AND LONDON

First published 2021
by Routledge
605 Third Avenue, New York, NY 10017

and by Routledge
2 Park Square, Milton Park, Abingdon, Oxon, OX14 4RN

Routledge is an imprint of the Taylor & Francis Group, an informa business

Library of Congress Cataloging-in-Publication Data
A catalog record for this book has been requested

ISBN: 978-0-367-62333-3 (hbk)

Typeset in Sabon
by Apex CoVantage, LLC

Contents

Figures

Tables

About the Contributors

Fiona Eva Bakas is a critical tourism researcher with international teaching experience. She holds a PhD in tourism (Otago University, 2014), has 20 years of varied work experience (corporate and academic), and is currently a contracted postdoctoral researcher in a nationwide project on creative tourism in rural areas and small cities (CREATOUR) at the Centre for Social Studies of the University of Coimbra (Portugal). Fiona is a member of research groups: CCArq (Coimbra), GOVCOPP (Aveiro), and ETEM (University of the Aegean). Her research interests include creative and cultural tourism, gender in tourism labor, qualitative methodologies, cultural mapping, handicrafts, entrepreneurship, rural tourism, and ecotourism.

Virginia Barbarossa is a doctoral student in economics, management, and quantitative methods in the Department of Management and Economics of the University of Salento, Lecce (Italy). She graduated with honors in business management at the same university. She has published an article on the role of language in slow tourism communication in a peer-reviewed scholarly journal and a contribution on slow tourism in a conference's proceedings. Her current research interests focus on word of mouth and the role of emotion in consumer behavior, with applications in cultural events and attractions.

Buke Bashuna is a practicing entrepreneur in tour operations (tour packaging, tour guiding, airfares, and ticketing) based in Nairobi, Kenya. She has more than ten years of experience in the travel and tourism industry and holds a Master of Philosophy degree in tourism management and a bachelor's degree in tourism management, both from Moi University, Kenya.

Federica Buffa is Associate Professor of Economics and Business Management at the Department of Economics and Management, University of Trento (Italy). She is a member of the Emasus research group (*Economics, Management, and Sustainable Consumption*), University of Trento. Actively involved in economics and business management studies, her research activities focus on destination management, strategy and management of small and medium tourism enterprises (SMTEs), sustainable tourism, and relationships and networks in community destinations. Recent research is focused on the sustainability management of multi-stakeholder territorial systems, with a particular emphasis on stakeholder engagement and networks in community destinations, and on the environmental management practices of SMTEs. She is author and co-author of scientific contributions in national and international publications. She is lecturer of the master course "Destination Management" (Master program MaST, *Sustainability Management and Tourism*) and a member of the PhD program SUSTEEMS (*Sustainability: Economics, Environment, Management, and Society*).

Daniel P. Bumblauskas is Associate Professor of Management and Hamilton/ESP International Fellow of Supply Chain and Logistics Management at the University of Northern Iowa (United States); Vice President of PFC Services, Inc., a consulting firm based in Marietta, Georgia; and a visiting professor at the University of Washington. He holds a courtesy appointment at the University of Missouri where he previously held a faculty appointment. Dan conducts research, teaches, and consults on various areas related to operational excellence and business development. Dan has published more than 45 peer-reviewed journal articles and conference proceedings, including publications in journals such as *Expert Systems with Applications*, *IEEE Transactions on Industry Applications*, *International Journal of Information Management*, and *Business Process Management*. Dan earned his PhD in industrial engineering from Iowa State University. Before his faculty appointments, Dr. Bumblauskas was previously employed in industry by ABB Inc. and Sears Holding Corporation.

Verdiana Chieffi is a doctoral student in economics, management, and quantitative methods at the Department of Management and Economics of the University of Salento, Lecce (Italy). She has been adjunct professor at the East China University of Political Science and Law, Shanghai (China); research fellow at the L'Orientale University, Naples (Italy); and research collaborator at the X.ITE Research Centre of LUISS University, Rome (Italy). She has published a contribution on slow tourism in a conference's proceedings. Her main areas of research include collecting behavior in art and information dissemination in contemporary art. She is currently investigating the role of art collectors as art knowledge disseminators within a scientific research project in collaboration with the National Gallery of Modern and Contemporary Art of Rome.

Giovanna Del Gaudio holds a PhD in business administration from the University Federico II, Naples (Italy). She is author of a book on value creation and value capture within the airline industry; she has also written book chapters and articles for national and international journals in the fields of marketing, tourism, focusing on the application of the resource-based view, and innovation in the tourism industry. She has attended several national and international conferences, where she has contributed co-authored papers. Her research focuses on value creation and value capture, dynamic capabilities, strategic destination governance, destination management and marketing, innovative marketing, concentrating on the application of the resource-based view, and open and user innovation perspectives in the tourism industry.

Valentina Della Corte is Full Professor of Business Management at Federico II University, Naples (Italy). She is Academic Coordinator of the Hospitality Management Course at Federico II University and teaches fundamentals of management, management and marketing of hospitality firms, tourism business management, and strategic management and marketing policies. Editor of international journals, she has authored numerous publications in journals, including *Tourism Management*, *International Journal of Tourism Research*, *British Food Journal*, *European Journal of Innovation Management*, and *Sinergie Italian Journal of Management*; she also wrote books as well as book chapters edited by Elgar, IGI Global, and McGraw Hill. She has participated—as chair, speaker, and reviewer—to national and international conferences such as the Strategic Management Society Annual Meeting, Academy of Management Annual Meeting, Italian Academy of Management Annual Meeting (AIDEA), Global Conference on Services, Society of Open Innovation, Technology and Martes Annual Meeting, and Sinergie Annual Meeting. Her research focuses on strategic management and marketing and, in particular, resource-based theory, business networks and coopetition, dynamic capabilities, value creation and appropriation both intra-firm and inter-firm, innovative marketing in tourism, and heritages and agri-food business.

Frédéric Dimanche is Professor and the Director of the Ted Rogers School of Hospitality and Tourism Management at Ryerson University, Toronto (Canada). He specializes in marketing and management of tourism organizations and destinations, and he has published in leading journals such as *Annals of Tourism Research*, *Tourism Management*, *Journal of Sustainable Tourism*, and *Journal of Travel Research*. An active lecturer and researcher, he has contributed to programs and universities, as well as to national or regional organizations, in Africa, Asia, Europe, and North and Central America. A member of the Travel and Tourism Research Association (TTRA) since 1990, he served as president of the *TTRA South Central Chapter* (United States) and president of *TTRA Europe*. He is currently a board member of *TTRA Canada* and of *Tourism HR Canada*. He published a book in

French on hospitality management and another, in English, on tourism management in Russia. He was elected in 2013 to the International Academy for the Study of Tourism.

Nancy Duxbury, PhD, is Senior Researcher and Co-coordinator of the Cities, Cultures and Architecture Research Group at the Centre for Social Studies, University of Coimbra (Portugal). She is Principal Investigator of CREATOUR (2016–2020), a national research-and-application project that aims to catalyze and develop creative tourism for community benefit in small cities and rural areas across four regions of Portugal. Other areas of research include culture in local sustainable development; culture-based development models in smaller communities; and cultural mapping, which bridges academic inquiry, community practice, and artistic approaches to understand and articulate place-based meanings, heritages, and attachments. She is a member of the European Expert Network on Culture, Adjunct Professor of the School of Communication, Simon Fraser University (Canada); and Adjunct Professor in the Department of Journalism, Communication, and New Media, Thompson Rivers University (Canada).

Massimiliano Fissore is a doctoral student in economics and management of innovation and sustainability in the Department of Economics and Management of the University of Parma (Italy). He graduated cum laude at the University of Turin in business management, marketing and strategies; his master's thesis, titled "*Heritage and Territory as Stronghold Against Globalization: the Case of the Province of Cuneo*", received a recommendation for publication. His main areas of research include cultural and territorial development, smart cities, and local identities.

Mariangela Franch is Full Professor of Marketing at the University of Trento (Italy). She founded the interdisciplinary Emasus research group (*Economics, Management, and Sustainable Consumption*), which brings competencies in tourism management, marketing, environmental economics, information and communication technologies, and statistics to the analysis of sustainable tourism. She is coordinator of the Master program MaST (*Sustainability Management and Tourism*) and a member of the PhD program SUSTEEMS (*Sustainability: Economics, Environment, Management, and Society*) at the University of Trento. She is a member of the International Association of Scientific Experts in Tourism (AIEST). Her main areas of research focus on decision-making processes, behavioral profiles of tourists, and sustainability management. She has authored or co-authored more than 180 publications in national and international journals.

Gianluigi Guido holds a PhD from the University of Cambridge (UK) and currently is Full Professor of Business Management in the Department of Management and Economics of the University of Salento, Lecce (Italy), where he teaches marketing and place marketing. He has been Professor at the LUISS University of Rome, at the Faculty of Psychology of the University of Rome "Sapienza", and at the Faculty of Statistics of the University of Padua, Italy. He has been visiting researcher in the Department of Psychology at the University of Florida, Gainesville (United States), and visiting scholar at Stanford University, California (United States). He was director of the social sciences area of the ISUFI (Istituto Superiore Universitario di Formazione Interdisciplinare) of the University of Salento, president of the degree courses in management studies, and Rector's Deputy for Institutional Communications at the same university. He has published more than 250 articles in major refereed scholarly journals, 20 books, and more than 50 chapters in edited books.

Maria Hakkarainen is a senior lecturer in tourism studies at the University of Lapland, Multidimensional Tourism Institute (MTI), Finland. She is an expert in tourism participatory economies: tourism mobility ecosystems, sharing economy in tourism, participatory research methods, and project development. She conducts research, teaches, and trains within her areas of interest and also has authored and co-authored publications in tourism journals and book chapters. Hakkarainen has been involved in several national and regional tourism development processes and projects in Finland. Currently, she works as a principal investigator in the Travelers Hospitality and Mobility Ecosystem FIT ME! project.

Dennis Heaton is Professor of Management and Dean of the College of Business Administration at Maharishi International University (MIU) in Fairfield, Iowa (United States). His previous positions at MIU include Director of the PhD program in Management and Dean of Distance Education and International Programs. He received his EdD in educational leadership from Boston University. His publications cover sustainable development, responsible management education, and appreciative inquiry. He has directed more than a dozen doctoral dissertations in sustainable business. He is co-editor of *Consciousness-Based Education and Management*, published in 2011 by Maharishi University of Management Press. Dr. Heaton has extensive experience teaching and traveling in the People's Republic of China.

Daria Hołodnik is Assistant Professor of Tourism Management at the Opole University of Technology (Poland). Her main fields of research, teaching, and training encompass including business models, marketing, and consumer behavior in tourism, hospitality, and gastronomy, and network analysis of tourism destination; however, her core competences particularly regard wine business models, wine tourism, and wine routes development. In the eight years of her academic career, she has published more than 30 articles, one authored book, two co-authored books, and several book chapters. She actively participates in the Academy of Wine Business Research and takes part in international research projects (currently together with the Wine University from Rhone-Alp and Grenoble University). Her teachings have been appreciated on international scenes, including Malaysia, Morocco, the United States, Spain, and France. At her mother university, she belongs to the teaching staff at the bachelor's and master's levels in tourism and recreation, as well as in tourism management.

John S. Hull is Associate Professor of Tourism Management at Thompson Rivers University (Canada). His main areas of research and teaching address destination policy and planning in peripheral and rural regions. He has published on mountain tourism, wellness tourism, creative and indigenous tourism, and food and wine tourism. John is particularly interested in the resiliency, well-being, and sustainability of tourism destinations. He presently serves on the British Columbia Minister's Tourism Engagement Council and is a member of the International Competence Network for Tourism Research and Education (ICNT), TTRA Canada, the New Zealand Tourism Research Institute (NZTRI), the Sonnino Working Group (Italy), and the tourism research center at Harz University of Applied Sciences (Germany). He has co-edited two books on mountain tourism (2016) and on multi-stakeholder perspectives of the tourist experience (2018).

Salla Jutila is a university teacher and doctoral student at the University of Lapland, Multidimensional Tourism Institute (Finland). In addition to sharing economy in tourism, her research and teaching interests include sustainable tourism, inclusion in tourism, accessible tourism, and tourism foresight. She has authored and co-authored publications in Finnish tourism journals as well as several book chapters. Jutila has been involved in regional, national, and international tourism development projects and was a project manager for a project investigating possibilities and challenges of peer-to-peer accommodation.

Salil S. Kalghatgi is an account executive for Cook Medical Devices, where he partners with healthcare provider supply chains in Southern California and Hawai'i (United States) to build sourcing and procurement solutions of endovascular devices. Salil previously worked as an analyst with a decade of experience driving big data analytics and strategy for industry leaders in entertainment, telecom, advertising, and retail. He continues engaging in freelance consulting, with a vision to promote data-driven decision-making and has strong interests in teaching and researching business analytics, continuous improvement, government finance, and policy, having published in *Public Money &*

Management. Salil studied history at Grinnell College, Iowa (United States), holds a BA in accounting from the University of Northern Iowa, and lives in San Diego with his partner, Dara Mead, and their cat, Herbie. In his free time he enjoys making maple syrup with friends in Iowa and traveling.

Nehemiah Kiprutto is a senior lecturer in the Department of Tourism and Travel Operations Management at Moi University (Kenya). He has authored and co-authored numerous travel and tourism publications in leading journals. He teaches both undergraduate and postgraduate programs at the Department of Tourism and Travel Operations Management. He also supervises projects and theses at both levels.

Sabrina Latusi holds a PhD in business administration at the University of Naples "Parthenope" (Italy). She is Associate Professor of Management at the University of Parma. Her main areas of research, teaching, and training include retail marketing, channel management, service marketing, consumer behavior, place branding, and management. Within her areas of interest, she authored and co-authored publications in leading marketing journals, together with book chapters and two monographs. She is a member of the faculty of the PhD program in economics and management of innovation and sustainability jointly promoted by the University of Parma and the University of Ferrara.

Serena Lonardi has been a pre-doctoral research assistant at the University of Innsbruck, Austria, and a member of the doctoral program, tourism and leisure in mountain regions, since 2019. She holds an MA in linguistic mediation, tourism, and culture from the University of Trento (Italy). Her research interests address intangible heritage, especially minority languages, tourism experiences, sustainable tourism, and destination attractiveness. Her doctoral thesis analyzes the contribution of minority languages in differentiating the destination and consequent role of the tourism field in the process of language preservation. The focus of the thesis is on two minority communities in northeastern Italy: the Cimbrian community in Luserna, near Trento, and the Ladin community in South Tyrol. Her research has recently been published by *Annals of Tourism Research* and has been presented at various conferences, including the TNC Conference 2018 in Kaprun (Austria) and Languages and Territory Conference 2019 in Trento (Italy).

Umberto Martini is Full Professor of Economics and Business Management at the University of Trento (Italy) Department of Economics and Management. He is currently teaching courses in tourism and territorial marketing, advanced marketing, and marketing of culture and tourism (the last at the Department of Humanities in the master program in linguistic mediation, tourism, and culture). At the Trentino

School of Management, he was director of a first-level Master program, Tourism Management, and was also responsible for the teaching area of destination management. The research focus has primarily been about management and marketing of tourist destinations, with a particular focus on the issue of sustainability. He coordinated the EU-funded research project (part of the Central Europe Programme), "Listen to the Voice of Villages" (LISTEN), and from March 2010 to November 2012, together with Prof. Mariangela Franch, the AIDEA (Italian Academy of *Economia Aziendale*) Study Group, "Management for Sustainable Development and Destination Competitiveness". Currently, he is the coordinator of the National Cluster on Management of Tourism and Territory at CUEIM (Consortium of Universities in Industrial Economics and Management, Italy). Within his areas of interest, he authored and co-authored publications in leading tourism journals, together with book chapters and monographs. He is also referee for many national and international scientific journals in the field of tourism and management.

Eleonora Moratti graduated in economics and management, receiving her master's degree in sustainability management and tourism (Master MaST) at the University of Trento (Italy). After completing an internship at a communications agency (HMC—Haas Marketing & Communications) for tourist destinations in Innsbruck, she worked in Effegilab as a marketing assistant and as a marketing manager at tooly.tips, a start-up company in the hospitality-tech sector. Strongly fond of web marketing and digital strategies in the tourism sector, over the past year she has developed copywriting skills in Xplace, a digital communications agency.

Pier Luigi Novi Inverardi is Associate Professor in Statistics at the University of Trento (Italy). His current research interests concern the broad area of mathematical statistics and are mainly focused on the concept of entropy and its inferential use in estimating probability distributions through Jaynes' Maximum Entropy principle constrained by fractional and/or translated moments, on robust multivariate inferential procedures and outlier detection. Over the years, he has gradually established an intense collaboration with the Emasus research group at the Department of Economics and Management (University of Trento), providing the quantitative skills needed for sample surveys and related data analysis, for reconstruction of the behavioral and organizational profiles of tourists and hoteliers of the Alpine area, paying particular attention to the sustainability of managerial choices. He is the author of numerous scientific publications, both for national and international journals, and he is doing referee activity for national and international scientific journals. He is a member of the Italian Statistical Association, the Association for Mathematics Applied to

Social and Economic Sciences, and the European Mathematical Association. He teaches undergraduate, master's, and PhD courses both at the Department of Economics and Management and at the Department of Mathematics of the University of Trento.

Petra Paloniemi is Senior Lecturer of Tourism at the Lapland University of Applied Sciences and a doctoral student at the University of Lapland, Multidimensional Tourism Institute (Finland). Her main areas of research and teaching include emergent trends in tourism such as value creation in sharing economy, sustainable development, experience design, and digital marketing. Within her areas of interest she has co-authored publications in some tourism-related journals. Petra has been involved in regional, national, and international tourism development projects, for instance, in a project investigating possibilities and challenges of peer-to-peer accommodation. She is a member of The Finnish Society for Tourism Research and ICRT Finland (International Centre for Responsible Tourism Finland Network).

Cecilia Pasquinelli, PhD, is Assistant Professor of Business Management at the University of Naples "Parthenope". Previously, she worked as a researcher at the Gran Sasso Science Institute, Italy, and at the University of Uppsala, Sweden. Her research has been mainly focused on place branding, place marketing, urban tourism, and local and regional development.

Alessandro M. Peluso is Associate Professor of Business Management in the Department of Management and Economics of the University of Salento, Lecce (Italy), where he has been teaching business management, marketing, marketing research, tourist destination marketing, and place marketing. He has been Adjunct Professor of Advertising, Marketing, and Marketing Strategies at the LUISS University, Rome (Italy) and has taught marketing research, customer analytics, and big data for marketing at the LUISS Business School. His research activity focuses on consumer behavior, sustainable consumption, and social communication. He has published three books and more than 50 scientific articles in peer-reviewed journals, such as *European Journal of Marketing*, *International Journal of Advertising*, *International Journal of Research in Marketing*, *Journal of Business Ethics*, *Journal of Business Research, Journal of Consumer Research*, *Journal of Marketing Research*, *Research Policy*, and many others, in addition to several contributions in book chapters and conference proceedings.

Kazimierz Perechuda is Full Professor of Business Management and Business Informatics at the Wroclaw University of Economics and Business (Poland). His main areas of research, teaching, and training include corporate business models, e-business models, modern management methods, digital marketing, virtual enterprise, agile firm, networking corporations, knowledge management, information management, process management, benchmarking, outsourcing, user experience, coaching, smart cities, tourism and hospitality management, agriculture tourism, vine rural tourism, and methodology of management sciences. He is the author and co-author of more than 300 scientific research publications, including more than 20 books, which were published in leading management journals and business publishing houses. He is the head of the Department of Business Intelligence in Management at the Faculty of Management, Wroclaw University of Economics and Business. He formerly was the head of the Organization and Management Department at the University of Physical Education in Wroclaw (Poland) and professor at Technical University in Zielona Gora (Poland). He was the former rector of the School of Management and Entrepreneurship in Walbrzych (Poland). Kazimierz is a member of the Scientific Society of Organization and Leadership (Poland) and Scientific Society of Business Informatics (Poland). He is a former member of the Polish Academy of Sciences Committee of Organization and Management Sciences (2003–2014). He has supervised eight PhD theses and served as teaching staff for PhD studies in management at various Polish universities.

Fabiana Sciarelli is Associate Professor of Management at the University of Naples 'L'Orientale'. At the same university, she is also a member of the planning and fund-raising commission; delegate spin-off and start-up for the university; a member of the Research Commission for the DSLL; and referent of the research titled "Management of Education in the Weak Areas of the World". She works as a project coordinator, teacher trainer, and researcher for many public and private companies in the field of non-profit, art, and culture. Fabiana is the author of numerous publications in journals such as *Land Use Policy*, *Sustainability*, *Economia della Cultura*, *Sinergie Journal*, *Il Capitale Culturale*, and *Journal of Systemics, Cybernetics and Informatics* (JSCI). She also wrote books as well as book chapters edited by Palgrave MacMillian, Electa, Franco Angeli, Marsilio, and McGraw Hill. She has participated—as chair, speaker, and reviewer—at national and international conferences about tourism, culture, and human development. She is currently involved in the project management of an international proposal that seeks to link research, innovation, art, and schools. She has been the director for many cultural and research projects, and she has been teaching for more than 15 years.

Simon M. Thiong'o is a Commonwealth Shared Scholar and MSc (wildlife conservation) degree holder from the University of Chester (UK). He also holds a bachelor's degree with first-class honors in tourism management from Moi University (Kenya). His research interest areas are protected area management and governance, biodiversity conservation, community-based tourism, and tourism sustainability, among others under which he has published several papers, teaching materials, and conference proceedings. He currently serves as a research assistant at the Department of Tourism and Travel Operations Management, Moi University, and teaches several undergraduate programs, including destination management, tourism and the environment, project management in tourism, and contemporary issues in travel and tourism, among others. He also supervises undergraduate projects and helps in postgraduate program (masters and PhD) administration and coordination in the department. Simon is a registered doctoral student (responsible tourism) at Leeds Beckett University (UK).

Bob Wishitemi is Professor of Tourism and Wildlife Studies in the Department of Tourism and Travel Operations Management, Moi University (Kenya). His areas of research/teaching include wildlife conservation and management, protected area management, community-based conservation initiatives, wildlife-based tourism, and pro-poor tourism. He has been in tertiary/university teaching and research for more than 30 years and has served in several administrative and academic positions, including deputy vice-chancellor—Academics, Research, and Extension at Moi University; principal, Kenya Wildlife Service Training Institute (Naivasha); head of several departments; and dean, Schools of Natural Resource Management and Business and Economics at Moi University. He has authored/co-authored several book chapters, conference proceedings, and more than 60 research papers in referred journals. Bob has been a visiting professor at Wageningen University (the Netherlands) and co-chaired the World Conservation Union's (IUCN) Task Force on Category V. He currently serves as the faculty academic reader, postgraduate programs coordinator, and masters and doctoral research supervisor.

Shuhui Xing conducted her doctoral dissertation research at Maharishi International University in Fairfield, Iowa (United States). The site of her research was the Bing Lang Gu ethnic tourism site near Sanya, Hainan Province, China—her hometown. Before her MBA and PhD studies in the United States, she earned her bachelor's degree in hotel management from the University of Science and Technology in Macau, China, in 2013. She also spent time at Florida State University in Tallahassee as an exchange student studying hospitality, and she completed a six-month internship at Disney World in Orlando, Florida.

Preface

The edited book "Humanistic Management and Sustainable Tourism: Human, Social and Environmental Challenges" is part of a two-volume collection, including also "Humanistic Tourism: Values, Norms and Dignity". It is aimed at contributing knowledge to tourism management from the broad perspective of humanistic management, thus offering a novel and original approach that has been neglected so far. As humanistic management is centered on the value of human life, the protection of human dignity, and the promotion of well-being, while still aiming for prosperity of the economy and of society, this approach is compatible with the founding concepts of tourism sustainability. A critical exploration of the nexus between humanistic management and sustainable tourism is not only consistent but is also important and urgent, from both a theoretical and managerial perspective.

Tourism is a fast-growing and changing industry and has become a driver of economic development in both developed and developing countries. As such, it raises concerns about the environmental and social pressures it causes. The tourism industry's potential for shared value creation and sustainable development is acknowledged. However, it remains a challenge for businesses, organizations, and tourist destinations. This is because sustainable tourism arguably conflicts with the predominant neoliberal structure of the economy and with hierarchical, profit- and consumption-driven societies. The emphasis on competition, growth, and profitability may undermine economic viability by consuming unreproducible resources and by undermining equity, justice, fairness, decent work, and social interaction. These principles/values have to do with the six 'essential elements'—dignity, people, prosperity, social justice, planet, and partnerships—that are considered requirements for the achievement of the 2030 Agenda for Sustainable Development (UN, 2015). They conceptually link sustainable development with humanistic management. This nexus is evident, among others, in sustainable development goal 8, which combines the International Labor Organization (ILO)'s "decent work agenda with economic growth" with the goal of achieving "sustained, inclusive and sustainable economic growth, full and productive employment and decent work for all" (UN, 2015, p. 23).

The crises recurrently challenging global travel and tourism, including the COVID-19 pandemic, climate change, bushfires, and extreme weather disasters—among others—show the weaknesses of neoliberal approaches and the collective economic dependence of countries on a form of tourism that is, at the least, very vulnerable, if not completely unsustainable. This requires the realization that our collective future depends on developing new approaches and interpretations of tourism, to effectively respond to human, societal, social, and climate challenges.

This book studies whether and how tourist destinations, organizations, and businesses are managed to create shared value and foster sustainable development, and whether these managerial practices follow humanistic management. In so doing, it focuses on the interplay among people/tourists and economic actors, the business organizational context and the wider socio-economic and cultural contexts. This inevitably involves multiple theoretical approaches, methods, and practical cases, on an international scale. For these reasons, the book may be of interest to a wide variety of readers. These include tourism scholars, researchers, and graduate students of many sub-disciplines of management and organization research, along with practitioners interested in finding ways to reset and rebuild tourism in order to manage it in a more humanistic and sustainable way.

Implicitly and explicitly, these ways also respond to the current urgency to implement strategies for recovery from the worldwide impact of the pandemic crisis and to provide a vision of what tourism could and should be when it recovers.

The edited volume aims at the following:

- Developing awareness of alternative routes for shared value creation in tourism, grounded in sustainability and the recognition of the value of human life—and their mutual reinforcement, that is, *humanistic tourism*;
- Offering approaches and tools for shared value creation based on collaboration among stakeholders in strategy design and the innovation of business models both levering on the integration of economic, social, environmental, and ethical values in their vision and/or value proposition; and
- Understanding to what extent cultural context matters in shaping approaches and tools for shared value creation and sustainable tourism from both the supply and demand side.

Book Contents and Structure

The book contains 14 contributions, organized into three separate parts, and a conclusion section, covering various aspects of the link between humanistic management and sustainable tourism. The opportunities for nurturing this convergence to foster a profound change in tourism and the economy are framed by the editors in the opening chapter, "The Capitalist Dilemma in Modern Tourism Development". Maria Della Lucia and Ernestina Giudici analyze the predominant neoliberal structure of capitalism and the vital function of tourism in sustaining it. Amid the pandemic crisis, paradoxes, and tensions of the capitalism and tourism systems urgently require rethinking. Shared value creation has been striving to overcome these vulnerabilities and paradoxes by interconnecting sustainability, social responsibility, and humanistic management. Thus, the nature (economic, social, and environmental) and sources of value creation are multiplied. Challenges and contradictions require long-term resolution through systems change by value co-creation dynamics.

Part I, "Stakeholder Cooperation for Socio-economic Development", incorporates three chapters devoted to the traditional—but evolving and still challenging—topic of engaging stakeholders in decision-making and activities aimed at sustainable development. The disparate viewpoints, roles, and power of players are further complicated by regulatory requirements, societal goals, new socio-economic production and consumption mechanisms, and technological advancement—among others.

Chapter 2, "The Seven Sisters of Piedmont: a Case Study of Potentially Worthwhile Cooperation", written by Sabrina Latusi and Massimiliano Fissore, analyzes the advantages and difficulties of cooperation among small-scale cities in the development of so-called slow tourism, through the case of the Seven Sisters of Piedmont, Italy (seven small cities). Results reveal tensions between institutional stakeholders (who are against) and non-institutional stakeholders (who are in favor) regarding the development of a partnership among these cities and the network's positioning in relation to slow tourism as a form of humanistic tourism. Administrative boundaries, turf rivalries, and parochialism are disincentives to policy makers for participating in this network-based project.

Chapter 3, "Do Socially Responsible Tour Operators Improve Tourism Sustainability? The Case of Nairobi, Kenya", co-authored by Simon M. Thiong'o, Buke Bashuna, Bob E.L. Wishitemi, and Nehemiah Kiprutto, sheds light on the role of tour operators in fostering sustainable development in Africa by adopting social responsibility initiatives. Results reveal that the perceived high importance of certain stakeholders (the tour operators) encourages socially responsible behavior. This results in participation in rehabilitation of tourist attraction sites, with the aim of not only giving back to society in terms of long-term development but also in terms of reputation and competitive advantage.

Written by Petra Paloniemi, Salla Jutila, and Maria Hakkarainen, Chapter 4, "Sharing Economy in Peripheral Tourism Destinations: the Case of Finnish Lapland", shows how the positive impact of a sharing economy has radically changed the tourism and hospitality business even in peripheral tourism destinations, as in the case of Finnish Lapland in the Arctic Circle. It offers the chance for individuals to participate in tourism, find employment, gain self-confidence in running a business, and meet new people. These circumstances have the potential of enhancing personal dignity and bringing humanism to business. However, planning sustainable destination management strategies is needed to enhance common well-being at the destination and to avoid conflicts among different stakeholders.

For exploiting the full potential of engaging stakeholders and society in the place/organization value chain, Part II introduces "New Business Models for Creating Shared Value". Four chapters deal with the alignment and integration of social, ethical, and environmental concerns/goals into a business model, in order to be successful in achieving sustainability. This incorporation challenges the interpretation and enunciation of value pillars (proposition, creation, delivery, and capture). It also shapes new forms of tourism and ways of engaging tourists. They combine the hedonic and eudaimonic dimensions of well-being, offering experiences both focused on pleasure and on meaningful and valuable activities aimed at the fulfillment of human potential.

In Chapter 5, Daria Hołodnik and Kazimierz Perechuda show how "The Application of Slow Life Coaching into an Agritourism Business Model" innovates the traditional agritourism farm and creates a form of creative tourism engaging customers in activities reinforcing the body-mind energy in different areas of life. This business transformation is fostered by the ability of slow life coaching to cultivate true human values and balance, combined with the extension of value co-creating for a local network of service facilities. The case of Lime Tree Valley Park, located in Poland, shows how slow life philosophy has been integrated successfully into the agritourism business model and coaching style. Education to slow life through creating an authentic experience is a way of applying the understanding of humanistic management to modern agritourism services.

Chapter 6, "Cultural Heritage Triggering Corporate Investments: 'Heritage Grab' or Sustainable Development?", written by Cecilia Pasquinelli, interprets private actors' dilemma in investing in cultural heritage through an analytical framework built by integrating the humanistic management perspective with land-grab literature. The dimensions of this framework—governance, business model, and access—are used to analyze three cases, Castelfalfi and Sammezzano in Tuscany, and Solomeo in Umbria, Italy, entailing private investment in cultural heritage as an asset for development. Solomeo, in particular, fits with a humanistic business model: legitimate and responsible corporate conduct pursues the vision of serving society and fostering citizens' quality of life, by involving the local community and attracting tourists to experience heritage and local culture mirrored into the entrepreneur's philosophy.

In Chapter 7, "Creative Tourism: A Humanistic Paradigm in Practice", Nancy Duxbury and Fiona Eva Bakas examine leading practices in contemporary creative tourism, focusing on rural and small cities. Eight initiatives by CREATOUR, a research-and-application project catalyzing a network of 40 creative-tourism initiatives in Portugal, provide case evidence on the ways in which creative-tourism strategies and practices embody and advance a humanistic paradigm. Results show that creative tourism promotes human flourishing, engages others in a journey of mutual discovery, honors the dignity of each stakeholder, and contributes to the common good in intriguing ways.

Another case of creative tourism and meaningful activity in the context of well-being is provided by Daniel P. Bumblauskas and Salil S. Kalghatgi in Chapter 8, "A Bloody Past With a Bright Future: Dark Tourism and Humanistic Management at Missouri State Penitentiary". The penitentiary is a historical landmark in Jefferson City, Missouri (United States), whose condition and aesthetics led to concerns in the community, creating a division between those in support of keeping the prison open for tourism and those in favor of demolishing it. Various aspects of dark tourism, service operations management, and humanistic tourism management are documented to distinguish between the motivations to restore or demolish landmarks and potential cultural heritage sites.

Part III, "The Influence of Cultural Context", comprises five chapters. They point out that sustainable and humanistic tourism in practice is, or should be, anchored to local identity, culture, and sub-cultures. Transformative actions that (re)value sustainability and well-being must take into consideration local values, norms, and ways of interaction. They should involve representatives of local culture and sub-cultures in decision-making regarding tourism development. Cultural characteristics matter not only in influencing the values that sustainable enterprises would like to create but also tourist behaviors.

Shuhui Xing and Dennis Heaton in Chapter 9, "A Grounded Theory Study of Ethnic Tourism in Hainan, China", present and discuss an ethnic tourism project in Bing Lang Gu, China, which has significantly achieved collaborative relationships with indigenous inhabitants whose culture and native land form the essence of an ethnic-minority theme park. This case illustrates the care taken in the inclusion and economic development of indigenous people while preserving their cultural heritage. This can occur when stakeholders share a common vision and are thus able to form collaborative relationships. The qualitative research undertaken to understand stakeholder relationships in the development of ethnic tourism involved five key stakeholder groups: local governments, tourism businesses, visitors, ethnic communities, and labor unions. Grounded theory analysis of the qualitative data led to the construction of an emergent human-centered theory of tourism development.

Another kind of ethnic tourism is proposed by Serena Lonardi, Umberto Martini, and John S. Hull in Chapter 10, "Positive Impacts of Sustainable Tourism: Preservation of Historical Languages and

Practices", analyzing the role of tourism in the preservation of endangered indigenous and minority languages in British Columbia, Canada, and Giazza, Italy, a Cimbrian community. Although there are considerable differences between the two cases, interviews with key informants showed that these minority cultures represent an added value. Tourists' interest led to an enhanced sense of pride among community members and a desire to learn more about their own traditional languages, as well as to make an effort to preserve their distinctive culture.

Chapter 11, "Spontaneous Tourism and Sustainable Development: The Evolution of the City of Naples", is positioned at the overlap between the supply and demand side. Fabiana Sciarelli, Valentina Della Corte, and Giovanna Del Gaudio inquire about the role of local community and other involved stakeholders in dealing with spontaneous tourism, which can lead to planned and systemic strategies also in the field of sustainability. This issue is analyzed in metropolitan cities by examining the case of Naples, Italy.

Unlike the previous contributions to Part II, which have a supply-side approach, the last two chapters adopt a demand-side perspective.

Chapter 12, "Everyday and Holiday Behaviors Regarding Sustainable Consumption Choices: Humanistic Management Perspectives for an International Destination", is co-authored by Mariangela Franch, Pier Luigi Novi Inverardi, Federica Buffa, and Eleonora Moratti. This chapter discusses how socio-cultural factors related to tourists' country of origin (COO) and their environmental awareness determine both the ways in which their holiday and everyday behaviors differ and different levels of willingness to pay (WTP) for more sustainable tourism services. The hypotheses were tested on a representative sample of international tourists around Lake Garda, Italy. Findings indicate a link between COO and virtuous holiday behaviors. The tourists' WTP totaled 20 million euros, which decision-makers could invest in key humanistic management objectives for the area, including environmental and cultural common goods.

In Chapter 13, "Memorable Experiences in Slow Tourism: an Empirical Investigation of Camping", Alessandro M. Peluso, Virginia Barbarossa, Verdiana Chieffi, and Gianluigi Guido focus on camping as a form of slow tourism that may deliver memorable experiences to tourists, activate internal transformative processes, and thereby contribute to human development. An empirical study carried out on a campsite in Apulia, in the south of Italy, shows that campers develop positive memories about experiences that convey a sense of renewal, connect with local people and culture, and are generally enjoyable. Such features increase tourists' satisfaction, propensity for sharing positive word of mouth, and likelihood of revisiting the campsite during a future holiday. The chapter concludes with a discussion of implications for scholars, managers, and policy-makers.

In Part IV ("Conclusions"), in Chapter 14, "Lessons for Shared Value Creation in Tourism: The Pandemic Challenge", Maria Della Lucia, Ernestina Giudici, and Frédéric Dimanche discuss the COVID-19 pandemic–induced impact on tourism in a critical perspective. The unprecedented societal changes brought on by the pandemic present an opportunity to set the ground for a transition toward alternative tourism development models. While there is still some wishful multidisciplinary thinking about the new tourism configuration, building what we call *humanistic tourism* revolves around processes of shared value creation at the interplay between sustainability and humanistic management. These processes entail four avenues for change—*human vs. human, human vs. nature, human vs. technology,* and *human vs. the economy*—which represent a possible alternative not only for tourism but also for managing businesses and fostering a better economy.

Reference

United Nations (UN). (2015). *Transforming our world: The 2030 agenda for sustainable development.* Retrieved from https://sustainabledevelopment.un.org/content/documents/21252030%20Agenda%20for%20Sustainable%20Development%20web.pdf

Acknowledgements

The editors are especially grateful to all the authors who, despite competing commitments, made this collaborative project possible. We also appreciate very much their cooperation in such a positive manner. Many thanks also to the reviewers for their valuable and constructive critiques: Albina Paskevic (Dalarna University, Sweden), Andrea Moretti (University of Udine, Italy), Angelo Presenza (University of Molise, Italy), Anke Winchenbach (University of Surrey, United Kingdom), Francesco Raggiotto (University of Udine, Italy), Greg Richards (Tilburg University, the Netherlands), Lara Penco (University of Genova, Italy), Marco Valeri (Università Nicolò Cusano, Italy), Mariapina Trunfio (Parthenope University, Naples, Italy), Michela Mason (University of Udine, Italy), Milena Viassone (University of Torino, Italy), Nicola Bellini (Sant'Anna School of Advances Studies, Pisa, Italy), Rachel Dodds (Ryerson University, Toronto, Canada), Rodolfo Baggio (Bocconi University, Italy), Sonya Graci (Ryerson University, Toronto, Canada), Tindara Abbate (University of Messina, Italy), Volker Rundshagen (Stralsund University, Germany), and Werner Gronau (Stralsund University, Germany). The editors and authors also express their appreciation to the rights holders of some of the illustrations used in this book, for their permission to reprint this material. Finally, we gratefully acknowledge Michael Pirson (Fordham University, New York), co-founder and president of the International Humanistic Management Association and editor of the Humanistic Management Series at Routledge, as well as the Routledge publishing staff, in particular Brianna Ascher and Naomi Round Cahalin, who were ever-present and supportive.

1 The Capitalist Dilemma in Modern Tourism Development

Maria Della Lucia and Ernestina Giudici

1. Opening Remarks

Capitalism has received a fair share of appreciation as well as criticism, and tourism has been identified to be vital for its sustainability (Fletcher, 2011). Tourism's predominant neoliberal structure is evident in its marketization and corporatization. These two factors have served the interests and profit-making agendas of the tourism industry's powerful players. Increasing international flows, mass tourism, and overtourism evident in many places manifest this approach as well as environmental impacts and concerns about livability of places and their loss of identity and culture (González Tirados, 2011).

Capitalism and tourism have shown a dilemma as a result of recurring crises. On the one hand, profits have become necessary to economic growth by increasing the number of tourists (the growth of the middle class in China and India). On the other hand, dynamic conservation and protection of human, natural, and cultural ecosystems has become crucial (Pung, Gnoth & Del Chiappa, 2020) to maintaining long-term sustainable development (Della Lucia, 2018). Therefore, these systems urgently require rethinking (Melè, 2016), but transition to alternatives remains an open debate.

This chapter discusses the foundation and contradictions of the economy and tourism's neoliberal structures and how their paradoxes are addressed through virtuous processes that multiply the nature and sources of value generation (Higgins-Desbiolles, Carnicelli, Krolikowski, Wijesinghe & Boluk, 2019; Pirson, 2019; Porter & Kramer, 2011, 2012). This chapter retraces the meaning and evolution of capitalism and further presents the relationship between capitalism and tourism and between tourism and sustainability. Finally, it introduces shared value as a notion that interconnects sustainability, social responsibility, and humanistic management and addresses capitalism and tourism challenges. The concluding section recommends a transition to new development models through these positive dynamics.

2. Capitalism: Meaning and Evolution

Capitalism has been the dominant economic system of Western economies since the end of World War II, evolving over the decades (Kazeroony, 2014). It is defined as a system based on individual rights, which serve as the basis for unethical behavior (Dettori & Giudici, 2014). Politically, legally, and economically, capitalism is a laissez-faire system (Scott, 1997), a system of objective laws (Bacher, 2007) and freedom applied to production, resulting in a free market (Phillips, 2013).

Batsch (2002) highlighted three phases of economic capitalism evolution. First, in "family capitalism, families own and manage companies". Second, under "managerial capitalism", management and ownership are separated (Berle & Means, 1932). Third, under "financial capitalism", economic and political domination is exercised by financial institutions rather than industrial capitalists (Gainet, 2014). This evolution reportedly emerged at the beginning of the 1980s in the United States.

Friedman (1962) strongly believed in free-market capitalism and argued that a business is mainly responsible for making profits for shareholders. In contrast, Freeman (1984) extended the scope of company objectives to stakeholders: "Any group or individual can affect or is affected by the achievement of the organization's objectives" (p. 46). Freeman further specified that "dividing the world into 'shareholder concerns' and 'stakeholder concerns' is roughly the logical equivalent of contrasting "apples" with "fruit". Shareholders are stakeholders, and it does not get us anywhere to try to contrast the two unless we have an ideological agenda that is served by doing so" (p. 46).

Capitalism has pros and cons. On the one hand, capitalism is seen to foster competitive advantage by stimulating entrepreneurship, innovation, and cost reduction in the product and labor market (Michael, 2014). On the other hand, capitalism has been identified to have endogenous irresponsibility (Johnston & Talbot, 2018), that is, it is less likely to generate responsible behavior (Streeck, 2016). "Capitalism only improves social and environmental conditions if this accidentally coincides with this goal" (Streeck, 2016, p. 2) as social behavior is subordinated to profit maximization.

Because of economic, financial, social, and environmental challenges, capitalism is faced with a dilemma that requires change. Its nature of being "irresponsible" needs to be transformed into a "responsible" system that is attentive to social responsibility and sustainability. This requirement does not indicate that capitalism has never changed. It has a dynamic, flexible nature and assumes different connotations (chameleon-like). However, these changes are more formal than substantial, such as those enunciated by Tomasi di Lampedusa in the book "The Leopard": "For everything to remain as it is, everything must change".

Capitalism needs substantial change as global challenges are deeply affecting its components: institutions, sectors, and actors in different roles (stakeholders, politicians, business owners, managers and employees, and academics and individual persons). Stakeholder interests must be addressed in a "sustainable and humanistic way" (Michael, 2014, p. 313).

The notion of "sustainable capitalism" (Gore & Blood, 2011) has been introduced to highlight the new connotation of the capitalist economic system given the ongoing crises and inequalities, and climate change threats. However, Tavanti (2014, p. 163) revealed that the "sustainable capitalism paradigm is a new and necessary solution which allows to maximize (only) long-term economic value creation". Sustainable capitalism may also provide adequate social (and environmental) responses by adopting a multi-stakeholder approach. In contrast, a shift of "focus from individual stakeholders to relationships among a range of stakeholders" (Martin, Roxas, Rivera & Gutierrez, 2020, p. 6) allows capitalism to create multidimensional values (Porter & Kramer, 2011, 2012).

A multi-stakeholder approach to sustainability emphasizes the capitalist system's complexity. This complex system can be interpreted by adopting a systems-thinking approach (Martin et al., 2020). Based on Von Bertalanffy's (1956) seminal study, many scholars have contributed to this approach and formulated various definitions, including Senge (1990); Forrester (1994); Richmond (1994); Sweeney & Sterman (2000); Stave & Hopper (2007); Kopainsky, Alessi & Davidsen (2011); and Squires, Wade, Dominick & Gelosh (2011). Arnold & Wade (2015) performed a synthesis of these studies and built on their experience to define systems thinking as "a set of synergetic analytic skills used to improve the capability of identifying and understanding systems, predicting their behaviors, and devising modifications to them in order to produce desired effects. These skills work together as a system" (p. 675).

The systems-thinking approach has been determined to capture sustainable capitalism complexity, that is, a large system of interdependencies among institutions, companies, and human beings. They are involved in addressing the world's economic, human, social, and environmental issues and evolve to a socially responsible and sustainable form. However, sustainable capitalism does not have a "human face". Thus, it "exploits" new connotations of social reality to pursue its objectives, increasingly becoming aware of environmental and social problems.

3. Capitalism and Tourism

Tourism marketization and corporatization over the past half-century is a function of capitalist expansion (Fletcher, 2011). Demand- and supply-side dynamics explain tourism's dramatic growth. From the supply side, interests and profit-making agendas of the tourism industry's powerful players have driven tourism growth. Mobile infrastructure development and innovation, and marketing and communication services and digitalization have fostered this trend. Governments, international organizations, and development planners have also expanded tourism as an economic growth strategy, especially in under-developed countries (Weaver, 2000). From the demand side, an increasing number of people with the desire and economic accessibility to travel, particularly the middle class, have nurtured tourism flows (MacCannell, 1999). Additionally, natural changes in industrial labor (paid vacation time, short working hours, less physically taxing jobs) and improved education have also allowed people to travel more.

Extending a systems-thinking approach to tourism interconnects capitalism and tourism (as a form of capitalism) in a single capitalist process. This process recalls the dynamics and challenges presented in the previous section. In other words, tourism's components interrelate and interact within the context of larger capitalistic systems. Moreover, the "capitalist world-system harnesses the tourism industry in order to sustain itself and attempt to resolve the internal contradictions of the present era" (Fletcher, 2011, p. 446).

This approach not only allows reframing tourism contradictions in a wider context but also examines sustainability and human dignity from the capitalism perspective. Tourism development fosters capitalism's continuous growth by way of finding outlets for other sectors' excess production, which may provoke an overproduction crisis. For example, tourism development requires manufacturing goods and industrial food in global value chains. Service workers in this production are often poorly paid, unskilled, and unsecured (Winchenbach, Hanna & Miller, 2019), thus extracting surplus value from workers' labor. Tourism expansion may also help address overtourism through displacement of capital from overdeveloped locations to newly developed ones. For example, spatial differentiation/distribution strategies may be adopted in solving these tourism issues (Haraldsson & Ólafsdóttir, 2018). These strategies can be accomplished also through international tourism development. In this perspective, sustainability reflects the internalization of natural (and cultural) resources as integral production conditions to be managed by capitalists to ensure their future exploitation (Fletcher, 2011). Anthropogenic climate change mitigation is part of the same phenomenon.

4. Tourism and Sustainability

The core argument in sustainable tourism primarily focuses on the interlink among development, governance, and management models (Della Lucia, 2018). Tourism development is shaped by norms (Weaver, 2012), namely, pro-growth and sustainability-conducive regulation, and hybrid norms, connected to stakeholder participation forms, which are deemed coercive, induced, and spontaneous (Tosun, 2006). These forms of participation empower players and interests in decision-making processes and related managerial strategies. Thus, these processes foster (or impede) shared value creation (Porter & Kramer, 2006, 2011), connecting (or disjoining) organizational success with social and environmental well-being over time.

The interplay among pro-growth development, top-down governance approaches, and managerial tools shapes the enduring agency and profit-making agendas of the tourism industry and governments' powerful players amid social and environmental challenges. Hence, expanding overtourism and mobility beyond a capacity threshold reveals ethical and environmental implications of uneven systems (ATLAS, 2020). Communication and marketing (Zeng & Gerritsen, 2014; Zuboff & Maxmin, 2002) and transport's (Hernandez Luıs, 2008) technology-driven democratization have further exacerbated tourism and transport access of different social groups, particularly the middle classes. This situation is observed to add pressure to social and environmental issues.

On the contrary, sustainable development positively correlates with integrated and participatory governance models and managerial practices (Della Lucia & Franch, 2017). Shared governance (Borrini-Feyerabend et al., 2013) entails institutional mechanisms and/or (in)formal processes. Thus, a coordination mechanism shares tourism management authority and responsibility with several actors, who are entitled to participate in planning and implementing of initiatives.

This collaborative governance model involves either weak (indirect participation) or strong (direct participation and information and consultation on decisions) forms of stakeholder engagement that combines top-down and bottom-up management approaches in different measures. Under weak stakeholder engagement, shared governance empowers participation of stakeholders who had previously been marginalized (e.g. local communities and indigenous people) by powerful interest groups to the extent that they are informed and consulted on decision-making and actions. These powerful players may have pro- and anti-growth goals (Della Lucia, 2018), such as the static conservation of protected areas and natural World Heritage Sites (WHS) (Mose, 2007). In strong stakeholder engagement, intensive active participation is encouraged and reinforced by sustainability goals but readapted for additional growth (Polnyotee & Thadaniti, 2015).

Widening participation allows shared governance to combine different goals in management strategies: suitable economic development and employment opportunities, environmental preservation and protection, and local community prosperity (Sotomayor, Arroyo & Barbieri, 2019). Examples include innovative and dynamic conservation strategies for protected areas and WHS (Mose & Weixlbaumer, 2007) and sustainable mass tourism strategies of spatial differentiation (Weaver, 2012). These tourism strategies also include regulated protected areas and WHS within large mass-tourism destinations.

5. Shared Value: The Nexus Between Sustainability, Responsibility, and Human Dignity

Amid the pandemic crisis, the paradoxes and tensions in the dominant neoliberal capitalist context and tourism growth model are gaining new insights. Shared Value (SV) (Porter & Kramer, 2011, 2012) is a notion that captures the multifaceted nature and value sources that allow the interconnection of sustainability, social responsibility, and humanistic management in order to address the paradoxes of these complex systems.

By questioning the purpose of business (Handy, 2002) and the ability of capitalism to foster prosperity (Jackson, 2011), Porter & Kramer (2011, p. 64) redefined the mission of business as "creating economic value in a way that also creates value for society by addressing its needs and challenges". Shared Value identifies this redefined purpose. Companies can gain competitive advantage and address social progress by treating social and environmental challenges as business opportunities—and *responsibilities*—pursued through corporate strategies (Porter & Kramer, 2006, 2012).

Although SV is a new concept, it reportedly overlaps significantly with established concepts, such as corporate social responsibility (CSR) and (corporate) sustainability. Shared Value may be assimilated to the gradual shift of business (Pirson & Lawrence, 2010) and social responsibility toward strategic-oriented approaches (Farmaki, 2019) because of global challenges. In this approach, CSR is integrated into firms and value chains. As a result, the company and society are mutually benefited by reconciling different interests (Goodwin, Font & Aldrigui, 2012). Previous responsive approaches focused on businesses' responsibility for their impacts on social, environmental, and ethical issues, and integrate CSR into organizational strategies and operations (Camilleri, 2014) to enhance competitive advantage and minimize business activities' negative impacts.

The ecosystem of SV (Kramer & Pfitzer, 2016) links it to sustainability. It captures the variety of sources—the different stakeholders participating in firm ecosystems—and the multifaceted nature—economic, social, and environmental—of SV. Their combination shapes the *collective-impact efforts* that enhance value co-creation (economic) and reduce value co-destruction (environmental and social). The coordinated efforts of players—from businesses to government agencies and charitable organizations and community members—can mobilize resources and capabilities from many entities sharing social transformation costs and new economic opportunities arising from social progress.

Business model innovation—the development of new visions and new architectures of organizational value proposition, creation, delivery, and capture—effectively overcomes structural market barriers to simultaneously gain profit and benefits from the natural environment and society. Business models for sustainability (Boons & Lüdeke-Freund, 2013;

Rauter, Jonker & Beumgartner, 2015) innovate business model value pillars by embedding sustainability. For example, hybrid organizations (social enterprises) incorporate value systems and action logics of various social sectors into their business models, thus exhibiting qualities of non-profit and for-profit enterprises (Haigh, Walker, Bacq & Kickul, 2015).

A humanistic management approach may also innovate business models (Pirson, 2019). This approach examines traditional business purposes and capitalism's ability to foster prosperity by promoting unconditional human dignity as a core organizational goal (Spitzeck, 2011). In turn, this goal is also a necessary condition for societal well-being (Pirson, Martin & Parmar, 2017) and thus sustainability (Dettori & Floris, 2019). In strict connection with justice, equality, and autonomy, recognizing and respecting individual value involve different processes of value creation for organizational and social progress. These processes entail recognizing people's will, consent, and decision-making capacity. They require people's active participation in decision-making processes in order to obtain fair (and sustainable) outcomes. They make possible that people can benefit from basic resources and societal goods, participate in their integral and holistic growth and fulfillment, and interconnect with the environment and all living beings (Melè, 2016).

Recognizing and respecting individual value in uneven systems provide principles of morality (Waldron, 2013) and capacity building (Nussbaum, 2011). These factors allow people to overcome violations and "to do" and "to be". The unlatching protection and development of these capabilities are conceived as substantial freedoms "from" constraints/violations and "to" thrive, and conditions for moral rights, x and y well-being, and social justice. Developing capabilities requires government expenditure and economic and social rights allocation to all citizens (Nussbaum, 2011).

Combined with sustainability, these principles can inspire uneven systems transformation at different levels (Jacobson, 2009), from the organizational to socio-economic and political contexts. Moreover, they guide actions toward alternative models based on re-discovery of what it means to be human and what matters most to humans.

6. Concluding Remarks

Recalling Mai and Smith (2015), Martin et al. (2020) observed that "tourism is not simply an industry but rather an entire system – complex and dynamic" (p. 3) characterized by interdependent relationships (Baggio, 2008). Tourism growth has been seen as a capitalist world economy function that facilitates temporary resolution of its contradictions (Fletcher, 2011). The persistence of challenges and contradictions requires long-term resolutions through systems change.

In the perspective of a systems-thinking approach, a weak element can determine a system's weakness given that multilateral interdependencies link each part. By contrast, a strong factor strengthens the system, that is, the strengths of tourism may positively affect tourism structure and, in turn, affect economic capitalism, and vice versa. Sustainability; social responsibility; humanization of businesses, economy, and society; and human behaviors may trigger pervasive shared value generation processes and foster the transition to new tourism models (Walter, 2016).

References

Arnold, R. D., & Wade, J. P. (2015). A definition of systems thinking: A systems approach. *Procedia Computer Science*, *44*, 669–678.

ATLAS Tourism and Leisure Review. (2020). Tourism and the Corona crisis: Some ATLAS reflections. *ATLAS*, *2*, 1–95.

Bacher, C. (2007). *Capitalism, ethics and the paradox of self-exploitation*. Munich: GRIN Verlag.

Baggio, R. (2008). Symptoms of complexity in a tourism system. *Tourism Analysis*, *13*(1), 1–20.

Batsch, L. (2002). *Le capitalism financier*. Paris: La Découverte.

Berle, A., & Means, G. C. (1932). *The modern corporation and private property*. New York: Commerce Clearing House.

Boons, F., & Lüdeke-Freund, F. (2013). Business models for sustainable innovation: State-of-the-art and steps towards a research agenda. *Journal of Cleaner Production*, *45*, 9–19.

Borrini-Feyerabend, G., Dudley, N., Jaeger, T., Lassen, B., Pathak Broome, N., Phillips, A., & Sandwith, T. (2013). *Governance of protected areas: From understanding to action*. Best Practice Protected Area Guidelines Series No. 20. IUCN Protected Areas Programme.

Camilleri, M. (2014). Advancing the sustainable tourism agenda through strategic CSR perspectives. *Tourism Planning and Development*, *11*(1), 42–56.

Della Lucia, M. (2018). *Approccio manageriale allo sviluppo turistico sostenibile. Un framework interpretativo e di gestione integrato*. Milano: Franco Angeli.

Della Lucia, M., & Franch, M. (2017). The effects of local context on World Heritage Site management: The Dolomites Natural World Heritage Site, Italy. *Journal of Sustainable Tourism*, *25*(12), 1756–1775. https://doi.org/10.1080/09669582.2017.1316727

Dettori, A., & Floris, M. (2019). Sustainability, well-being, and happiness: A co-word analysis. *International Journal of Business and Social Science*, *10*(10), 29–38. https://doi.org/10.30845/ijbss.v10n10a5

Dettori, A., & Giudici, E. (2014). Is it possible to achieve sustainable capitalism by 2020? In H. Kazeroony & A. Stachowicz-Stanusch (Eds.), *Capitalism and the social relationship: An organizational perspective* (pp. 291–302). Palgrave Macmillan. https://doi.org/10.1057/9781137325709_18

Farmaki, A. (2019). Corporate social responsibility in hotels: A stakeholder approach. *International Journal of Contemporary Hospitality Management*, *31*(6), 2297–2320.

Fletcher, R. (2011). Sustaining tourism, sustaining capitalism? The tourism industry's role in global capitalist expansion. *Tourism Geographies*, *13*(3), 443–461.

Forrester, J. W. (1994). System dynamics, systems thinking, and soft OR. *System Dynamics Review*, *10*(2–3), 245–256.

Freeman, R. E. (1984). *Strategic management: A stakeholder approach*. Boston: Pitman.

Friedman, M. (1962). *Capitalism and freedom*. Chicago: University of Chicago Press.

Gainet, C. (2014). Socially responsible investment: How shareholders change their role within the capitalism paradigm. In H. Kazeroony & A. Stachowicz-Stanusch (Eds.), *Capitalism and the social relationship: An organizational perspective* (pp. 183–196). Palgrave Macmillan.

González Tirados, R. M. (2011). Half a century of mass tourism: Evolution and expectations. *The Service Industries Journal*, *31*(10), 1589–1601. https://doi.org/10.1080/02642069.2010.485639

Goodwin, H., Font, X., & Aldrigui, M. (2012). 6th international conference on responsible tourism in destination. *Revista Brasileira de Pesquisa em Turismo*, *6*(3), 398–402.

Gore, A., & Blood, D. (2011, December 14). A manifesto for sustainable capitalism: How businesses can embrace environmental, social and governance metrics. *The Wall Street Journal*. Retrieved from www.wsj.com/articles/SB10001424052970203430404577092682864215896

Haigh, N., Walker, J., Bacq, S., & Kickul, J. (2015). Hybrid organizations: Origins, strategies, impacts, and implications. *California Management Review*, *57*(3), 5–13.

Handy, C. (2002). What is a business for? *Harvard Business Review*, *80*(12), 49–56.

Haraldsson, H. V., & Ólafsdóttir, R. (2018). Evolution of tourism in natural destinations and dynamic sustainable thresholds over time. *Sustainability*, *10*(12), 4788. https://doi.org/10.3390/su10124788

Hernandez Luis, J. A. (2008). *El turismo demasas. Evoluci´on y perspectivas* [*Mass Tourism: Evolution and perspectives*]. Madrid: Sintesis.

Higgins-Desbiolles, F., Carnicelli, S., Krolikowski, C., Wijesinghe, G., & Boluk, K. (2019). Degrowing tourism: Rethinking tourism. *Journal of Sustainable Tourism*, *27*(12), 1926–1944. https://doi.org/10.1080/09669582.2019.1601732

Jackson, T. (2011). *Prosperity without growth: Economics for a finite planet*. London: Routledge.

Jacobson, N. (2009). A taxonomy of dignity: A grounded theory study. *BMC International Health and Human Rights*, *9*(1), 3.

Johnston, A., & Talbot, L. (2018). Why is modern capitalism irresponsible and what would make it more responsible? A company law perspective. *King's Law Journal*, *29*(1), 111–141. https://doi.org/10.1080/09615768.2018.1478201

Kazeroony, H. (2014). Capitalism and the social relationship: A contextual overview. In H. Kazeroony & A. Stachowicz-Stanusch (Eds.), *Capitalism and the social relationship: An organizational perspective* (pp. 3–15). London: Palgrave Macmillan.

Kopainsky, B., Alessi, S. M., & Davidsen, P. I. (2011). Measuring knowledge acquisition in dynamic decision making tasks. In J. M. Lyneis & G. P.

Richardson (Eds.), *The 29th International Conference of the System Dynamics Society* (pp. 1–31). System Dynamics Society.

Kramer, M. R., & Pfitzer, M. W. (2016). The ecosystem of shared value. *Harvard Business Review*, *94*(10), 80–89.

MacCannell, D. (1999). *The tourist: A new theory of the leisure class* (2nd ed.). Berkeley: University of California Press.

Mai, T., & Smith, C. (2015). Addressing the threats to tourism sustainability using systems thinking: A case study of Cat Ba island, Vietnam. *Journal of Sustainable Tourism*, *23*(10), 1504–1528.

Martin, F., Roxas, Y., Rivera, J. P. R., & Gutierrez, E. L. M. (2020). Framework for creating sustainable tourism using systems thinking. *Current Issues in Tourism*, *23*(3), 280–296. https://doi.org/10.1080/13683500.2018.1534805

Melé, D. (2016). Understanding humanistic management. *Humanistic Management Journal*, *1*(1), 33–55. https://doi.org/10.1007/s41463-016-0011-5

Michael, A. (2014). Capitalism at a crossroads: Unfulfilled expectations and future challenges. In H. Kazeroony & A. Stachowicz-Stanusch (Eds.), *Capitalism and the social relationship: An organizational perspective* (pp. 303–319). London: Palgrave Macmillan.

Mose, I. (2007). *Protected areas and regional development in Europe: Towards a new model for the 21st century*. Hampshire: Ashgate.

Mose, I., & Weixlbaumer, N. (2007). A new paradigm for protected areas in Europe? In I. Mose (Ed.), *Protected areas and regional development in Europe. Towards a new model for the 21st century* (pp. 3–19). Hampshire: Ashgate.

Nussbaum, M. (2011). *Creating capabilities. The human development approach*. Cambridge: Belknap.

Phillips, B. (2013, February 16). Capitalism is good in theory and in practice. *Capitalism Magazine*, 104–110. Retrieved from www.capitalismmagazine.com/2013/02/capitalism-is-good-in-theory-and-in-practice/#:~:text=Facebook-,Upon%20hearing%20an%20argument%20for%20capitalism%2C%20many%20respond%2C%20%E2%80%9CThat,both%20theory%20and%20in%20practice.&text=A%20theory%20that%20achieves%20something,not%20a%20very%20good%20theory

Pirson, M. A. (2019). A humanistic perspective for management theory: Protecting dignity and promoting well-being. *Journal of Business Ethics*, *159*, 1–19. https://doi.org/ 10.1007/s10551-017-3755-4

Pirson, M. A., & Lawrence, P. R. (2010). Humanism in business—Towards a paradigm shift? *Journal of Business Ethics*, *93*, 553–565. https://doi.org/10.1007/s10551-009-0239-1

Pirson, M. A., Martin, K., & Parmar, B. (2017). A humanistic perspective for management theory: Protecting dignity and promoting well-being. *Journal of Business Ethics*, *145*(1), 1–19. https://doi.org/10.1007/s10551-017-3755-4

Polnyotee, M., & Thadaniti, S. (2015). Community-based tourism: A strategy for sustainable tourism development of Patong Beach, Phuket island, Thailand. *Asian Social Science*, *11*(27), 90–98.

Porter, M., & Kramer, M. (2006). Strategy and society: The link between competitive advantage and corporate social responsibility. *Harvard Business Review*, *84*(12), 78–92.

Porter, M., & Kramer, M. (2011). Creating shared value. *Harvard Business Review*, *89*(1/2), 62–77.

Porter, M., & Kramer, M. (2012). Shared value: The bridge between corporate social responsibility and corporate strategy. In A. Schneider & R. Schmidpeter (Eds.), *Corporate Social Responsibility* (pp. 137–153). Berlin/Heidelberg: Springer.

Pung, J. M., Gnoth, J., & Del Chiappa, G. (2020). Tourist transformation: Towards a conceptual model. *Annals of Tourism Research*, *81*, 1–12. https://doi.org/10.1016/j.annals.2020.102885

Rauter, R., Jonker, J., & Beumgartner, R. J. (2015). Going one's own way: Drivers in developing business models for sustainability. *Journal of Cleaner Production*, *140*(1), 144–154. http://dx.doi.org/10.1016/j.jclepro.2015.04.104

Richmond, B. (1994). Systems thinking/system dynamics: Let's just get on with it. *System Dynamics Review*, *10*(2–3), 135–157. https://doi.org/10.1002/sdr.4260100204

Scott, J. (1997). *Corporate business and capitalist classes*. Oxford: Oxford University Press.

Senge, P. M. (1990). *The fifth discipline: The art and practice of the learning organization*. New York: Doubleday/Currency.

Sotomayor, S., Arroyo, C. G., & Barbieri, C. (2019). Tradition and modernity side-by-side: Experiential tourism among Quechua communities. *Journal of Tourism and Cultural Change*, *17*(4), 377–393. https://doi.org/10.1080/14766825.2019.1591683

Spitzeck, H. (2011). An integrated model of humanistic management. *Journal of Business Ethics*, *99*(1), 51–62. https://doi.org/0.1007/s10551-011-0748-6

Squires, A., Wade, J., Dominick, P., & Gelosh, D. (2011). *Building a competency taxonomy to guide experience acceleration of lead program systems engineers*. Hoboken, NJ: Stevens Institute of Technology.

Stave, K., & Hopper, M. (2007, July). What constitutes systems thinking? A proposed taxonomy. In J. Sterman (Ed.), *Proceedings of the 25th international conference of the system dynamics society* (pp. 1–24). System Dynamics Society. Retrieved from http://static.clexchange.org/ftp/conference/CLE_2010/CO2010-06Session5MeasureST.pdf

Streeck, W. (2016). *How will capitalism end?* London and New York: Verso.

Sweeney, L. B., & Sterman, J. D. (2000). Bathtub dynamics: Initial results of a systems thinking inventory. *System Dynamics Review*, *16*(4), 249–286.

Tavanti, M. (2014). Sustainable development capitalism: Changing paradigms and practices for a more viable, equitable, bearable, and just economic future for all. In A. Kazeroony & A. Stachowicz-Stanusch (Eds.), *Capitalism and the social relationship. An organizational perspective* (pp. 163–182). London: Palgrave Macmillan.

Tosun, C. (2006). Expected nature of community participation in tourism development. *Tourism Management*, *27*(3), 493–504.

Von Bertalanffy, L. (1956). General system theory. *General Systems*, *1*(1), 11–17.

Waldron, J. (2013). The paradoxes of dignity—About Michael Rosen, dignity: Its history and meaning (Harvard University Press, 2012.). *European Journal of Sociology*, *54*(3), 554–561. https://doi.org/10.1017/S0003975613000404

Walter, P. G. (2016). Catalysts for transformative learning in community-based ecotourism. *Current Issues in Tourism*, *19*(13), 1356–1371.

Weaver, D. B. (2000). A broad context model of destination development scenarios. *Tourism Management*, *21*, 217–224.

Weaver, D. B. (2012). Organic, incremental and induced paths to sustainable mass tourism convergence. *Tourism Management*, *33*(5), 1030–1037.

Winchenbach, A., Hanna, P., & Miller, G. (2019). Rethinking decent work: The value of dignity in tourism employment. *Journal of Sustainable Tourism*, *27*(7), 1026–1043. https://doi.org/10.1080/09669582.2019.1566346

Zeng, B., & Gerritsen, R. (2014). What do we know about social media in tourism? A review. *Tourism Management Perspectives*, *10*, 27–36.

Zuboff, S., & Maxmin, J. (2002). *The support economy—why corporations are failing individuals and the next episode of capitalism*. New York: Penguin.

Part I

Stakeholder Cooperation for Socio-economic Development

2 The Seven Sisters of Piedmont

A Case Study of Potentially Worthwhile Cooperation

Sabrina Latusi and Massimiliano Fissore

1. Introduction

In destination management debate, policy-makers and researchers are moving toward the concepts of sustainability, ethics and identity as important strategic elements in fostering and developing the well-being of stakeholders (e.g. Franch & Martini, 2013; Su, Huang, S. & Huang, J., 2018). Sustainability is based on a humanistic vision of management, whereby places with a value-based perspective can combine economic, social, and environmental goals.

This perspective is promoting interest in a slow way of life and driving the growth of slow food, slow cities and slow tourism (Heitmann, Robinson & Povey, 2011). Slow tourism is a new frontier in tourism and destination management, particularly challenging for places with a historical, cultural and environmental heritage wishing to generate well-being for their stakeholders (e.g. Fullagar, Markwell & Wilson, 2012; Lowry & Lee, 2016).

This kind of heritage and positioning usually characterizes small, underestimated, and unrecognized places. When such places act independently, their small size makes for difficulties in visibility and consumer recognition. Therefore, cooperation among places (small cities, villages, cultural and natural beauty spots, etc.) seems to be the best way to pursue sustainable growth. Indeed, cooperation would seem to make it possible to overcome difficulties in the management and development of small places.

However, the application of the slow philosophy in tourism development and destination management has been studied mostly at a city level, in relation to Cittàslow (e.g. Mayer & Knox, 2006; Nilsson, Svärd, Widarsson & Wirell, 2011; Ekinci, 2014), at a regional (de la Barre, 2012; Murayama & Parker, 2012) and national level (Serdane, 2019). Research on opportunities and challenges for using slow philosophy in tourism development at a network-based geographical scale (i.e. non-standard region) is lacking in the academic literature. Networking can make for a viable and sustainable way of emerging in the slow tourism context by focusing strategies on cooperation (e.g. Guercini & Tunisini, 2017; Buffa, Beritelli & Martini, 2018).

The purpose of the chapter is to fill up this gap and, to this end, the following central research question is posed:

> *To what extent is cooperation among small-scale cities a valuable way to pursue slow tourism development as a mean to enhance dignity and well-being for the tourists and the local communities?*

From this basis, specific sub-questions can be distinguished:

- What are the potential advantages of stakeholder cooperation and networking?
- What are the main obstacles to stakeholder cooperation and networking?

To address these issues, the study investigates an Italian local area known as 'Sette Sorelle' (Seven Sisters). These seven small cities (Alba, Bra, Cuneo, Fossano, Mondovì, Saluzzo, and Savigliano) in the province of Cuneo (in Piedmont) show a relatively low tourism demand. They have little-known natural and cultural elements that can represent great potential for a sustainable tourism development according to a slow philosophy. These municipalities share a cultural and historical common denominator that could support networking among them and their actors and facilitate the identification of common goals. These cities could grow together, improve their citizens' quality of life, and become a valid alternative to other more well-known nearby destinations (e.g. Monferrato). Therefore, the Seven Sisters case represents a suitable context to explore cooperation and networking management in slow tourism development.

The chapter is organized as follows. First, an introductory theoretical discussion regarding the main topics under consideration (cultural slow tourism, nonstandard regions, and cooperation) is undertaken. Next, the methodology is presented, including the research context of the Seven Sisters. The opinions of local stakeholders on the development of the Seven Sisters are then analyzed, highlighting potential advantages and pitfalls. The chapter concludes by summarizing the main empirical findings, underlining the contribution made to existing knowledge.

2. A Theoretical Humanistic Framework for Cultural Slow Tourism in Nonstandard Regions

Humanistic management is a people-oriented management, showing care for people flourishing and well-being (Melé, 2016). It is a concept of management that upholds unconditional human dignity within an economic context (Melé, 2003; Spitzeck, Amann, Pirson, Khan & Von

Kimakowitz, 2009). The concept calls for a new vision of business: "serving the societies in which business operates, increasing their citizens' quality of life" (Spitzeck, 2011, p. 51). This humanistic-oriented ethics pursuing dignity and well-being should be considered not only as a trend but also as a need for management (Dierksmeier, 2011). Re-thinking economics, business, and management from a humanistic perspective is particularly relevant in the tourism industry (Cynarski & Obodynski, 2004). This orientation calls for a tourism impact that socially, culturally, and environmentally is neither permanent, nor irreversible (Beech & Chadwick, 2006), allowing citizens (tourists, local communities, workers, etc.) a healthy, productive, and pleasant life.

In recent years, quality of life and well-being have become an important area of investigation in tourism research (Jepson, Stadler & Spencer, 2019). Indeed, there has been a shift in the value of tourism toward more subjective elements, such as sustainability, wellness, well-being, and quality of life (Perdue, Tyrrell & Uysal, 2010). According to Uysal, Sirgy, Woo and Kim (2016), systematic reviews of quality of life studies in tourism, tourists' experiences and activities contribute to positive affect in a range of life domains and have a significant effect both on tourists' overall life satisfaction and on well-being of residents of host communities too.

Recently, sustainability and personal/social well-being have become driving forces underpinning new forms of tourism, among which is slow tourism (Moore, 2012; Oh, Assaf & Baloglu, 2016).

2.1 Cultural Slow Tourism

The phenomenon of slow tourism needs to be understood in the sociocultural context of slow movement (Fullagar et al., 2012). The slow movement is a lifestyle revolution (Honorè, 2005), a change of mindsets and a re-evaluation of life priorities (Heitmann et al., 2011). The new priorities include a slower approach to life, that strives for simplicity, mindfulness, and embodied experiences to facilitate renewal, revitalization, and self-enrichment (Howard, 2012). In the context of tourism, these new priorities translate to engaging with local people, preserving cultural traditions and natural resources, and making meaningful connections with heritage, places, and environment (e.g. Gibson, Pratt & Movono, 2012).

Slow tourism is an umbrella term encompassing various tourism types (Serdane, 2019), such as responsible tourism (Timms & Conway, 2012) and ethical tourism (Clancy, 2015). Several authors also link slow tourism with sustainability (e.g. Sidali & de Obeso, 2018).

From a humanistic perspective, previous studies show that slow tourism has the potential to improve the quality of life for both inhabitants and tourists (Hatipoglu, 2014; Presenza, Abbate & Micera, 2015) and enhance local community's empowerment (Conway & Timms, 2012; Park & Kim, 2015). Slow tourism prioritizes social, cultural, and environmental well-being and sustainability along with quality of experiences (Mayer et al., 2006; Heitmann et al., 2011; Ekinci, 2014). According to Timms and Conway (2012, p. 398), the slow-tourism model "offers a more sustainable tourism product that is less alienated (and alienating), more culturally sensitive, authentic and a better-paced experience for hosts and tourists alike". It can foster a sense of place and assist in the forging of individuals' and groups' identities (Park & Kim, 2015).

Therefore, slow tourism has the potential to enhance dignity and well-being for the tourists but also for people involved in such industry and for local communities (Caffyn, 2012). In essence, slow tourism can be considered part of a humanistic practice.

In slow tourism, history, heritage, and culture play a significant role as drivers of local sustainable development and well-being (e.g. Santagata, 2002). These intangible assets are relevant to both internal and external stakeholders. They are the foundations of the sense of belonging and pride of the inhabitants. In the meantime, they are crucial to the narrative of the place (Jensen, 2007), contributing to the perception of the destination as unique and to the sense of authenticity experienced (Heitmann et al., 2011). Moreover, a virtuous circle may develop because local residents play an important role in creating an appealing atmosphere for tourists. In turn, tourists' feedback can lead to a stronger appreciation of local culture and heritage among local inhabitants, strengthening their sense of identity (Nilsson, Svard, Widarsson & Wirell, 2007). Therefore, the interaction among culture, identity, and image is relevant (Kavaratzis & Hatch, 2013) in a humanistic perspective.

Culture can support the processes of place identification, that is, *identification of* the place as such, *being identified as* belonging to the place, and *identification with* the place (Kalandides, 2012; Boisen, 2015). Culture can represent a unifying element promoting place attachment, place recognition, and distinctive place positioning, combining a set of diverse resources (monument and heritage sites, cultural events, cuisine, traditions, etc.) in a complex interaction of values, stories, emotions, expectations, and experiences.

In slow tourism, the focus is on quality rather than quantity. This does not mean fewer experiences (Heitmann et al., 2011) but a valuable variety of experiences. Therefore, the development of tourism using slow philosophy requires cooperation among local stakeholders to provide a comprehensive tourism offering (Beritelli, 2011). Nevertheless, studies on slow tourism from the supply-side perspective are limited and mostly focused on a city level (Serdane, 2019). Cooperation among different places remains a topic largely unexplored in the application of the slow philosophy in tourism development and destination management.

According to Park and Kim (2015), from a tourism destination perspective, slow tourism is particularly suitable for small-scale cities. Cooperation and networking among these realities can overcome the difficulties they encounter, when acting independently, in acquiring visibility and consumer recognition. This leads to the topic of nonstandard regions.

2.2 Nonstandard Regions and Cooperation

It is a shared opinion that a place exists as an entity if there are people who agree on its existence (Boisen, 2015). Such recognition is relative to a spatially defined area that may, or may not, be limited by administrative-based borders. In the latter case, a new geographical entity emerges, although not institutionally acknowledged, since the recognition of the place transcends the administrative-based borders of the individual territorial units that together—sharing an identity/image—form the place itself (Metzger, 2013). This creates what is known as a nonstandard region (e.g. Terlouw, 2009), something proper to the public realm but also a matter of concern to other economic and social actors (Lucarelli, 2018). This is because a number of local stakeholders can shape the identity of a nonstandard region.

Indeed, there has been a change in the distribution of responsibilities over different scalar levels of spatial authority (from supranational to municipal) and between the public and private sectors. This can be seen as a community empowerment: local communities acquire rights and power to collect resources for their needs, and rethink management to achieve autonomy and self-reliance and to maximize their quality of life (e.g. Scheyvens, 2002). Consequently, local authorities have become increasingly responsible for, and directly involved in, their own economies (e.g. Brenner, 2004) to secure the welfare and well-being of their inhabitants. This increased responsibility is a strong incentive for local authorities to understand interactions among places and to engage in cooperation, prioritizing collaborative strategies over competitive ones. The relational and collaborative space emerging from cooperation crosses the established administrative-based borders and makes these existing borders less important (Deas & Lord, 2006).

However, since nonstandard regions are composed of distinct political and administrative units, their sustainability and evolutionary dynamics cannot be taken for granted (Pasquinelli, 2015).

According to Boisen (2015), collaboration is a valuable issue, and the resulting nonstandard region becomes a better place if the individual territorial units involved share a common denominator relevant to their internal and external stakeholders. This common element could be socio-economic, cultural, or historical. It should strengthen the value proposition of the area, its distinctiveness, and identity, and it should support a vision for a shared future. "Although this theoretically makes sense, this is not always the case in practice" (Boisen, 2015, p. 19).

In some cases, these new regions are the result of incentives and funding schemes; in other circumstances, they are formed through coalitions of companies, institutions and public authorities in an area dominated by one or more strong business sectors. There are other cases that result from a relational and collaborative space chosen as a model to support regeneration plans. This can occur through re-positioning and re-branding strategies, according to criteria of effectiveness and efficiency. Outside these cases, since the network is not a region from an administrative perspective, political interest in the network project can be very weak (Boisen, 2015). In fact, "who wants to be a politician in a region that does not exist?" (Berg & Lofgren, 2000, p. 10 as cited by Pasquinelli, 2015).

3. Research Method

3.1 The Research Context

The geographical focus of the research are seven small-scale cities in the province of Cuneo, historically known as the 'Seven Sisters'. The province of Cuneo, also called 'La Granda', is located in the northwest of Italy and is characterized by the heterogeneous and varied nature of many of its attributes, notably the environment, culture, and gastronomy.

The territory is divided into three main parts, where the sister-cities are located: the Langhe and Roero hills, with Alba and Bra; the mountains area, with Cuneo (the county seat), Mondovì and Saluzzo; and the plain, with Savigliano and Fossano.

Each one of these seven cities has a unique identity and some specific quality that distinguishes it from the others. At the same time, the cities share a historical and cultural heritage and are tied by a common thread of shared values. Consequently, they are similar enough to be considered sisters.

This territory is an important gastronomic center, especially renowned for wine, meat, truffles, chocolate, and cheese. In particular, Alba and the surrounding Langhe hills have achieved international recognition. While mainly based on agriculture, the economy is also characterized by the presence of important industrial sectors, which are well spread on the local area, giving to every sister-city a business support.

Historically, the Cuneo region and its cities owe their origins to a common cultural element: the aristocratic heritage and the centuries as part of the Savoy dominion.

Nevertheless, the territory is characterized by an intrinsic heterogeneity. This element reinforces the sense of identity and reciprocal diversity of each sister-city and is even a ground for dualism inherited from the past.

There is a dualism in the origins of the cities and their historical alliances: Cuneo and Mondovì as free commons; Saluzzo, Fossano, and Bra as aristocratic feuds; Savigliano and Alba as minor centers. Central to these differences are the different alliances among families and regions (the Kingdom of France, Monferrato, the Papal State, and the Holy Roman Empire). Over time, these differences were largely reconciled when the province of Cuneo was acquired by the Sabaud State.

The characteristic territorial heterogeneity gave rise to an interesting cultural flowering that manifested itself in architectural and artistic works, dialects, local cuisine, etc., which fostered interchange among the Seven Sisters, leading to collective growth and local pride.

3.2 The Seven Sisters as a Potential Nonstandard Region

The Seven Sisters share a long history, but nowadays they could be seen as a dormant place. However, because a place can emerge as a result of social processes, "dormant places can be brought back to life if they are relevant enough to contemporary people" (Boisen, 2015, p. 15).

Following this approach, the Seven Sisters could become a nonstandard region if the seven municipalities were to engage in a cooperative approach to sustainable economic and social development. The cities could be partners in a network that blurs their administrative boundaries and creates a new relational and collaborative space. Therefore, the network could become a platform for imagining new strategic trajectories to conceive and implement local, sustainable development policies (Pasquinelli, 2015), fostering the well-being of the region and its inhabitants.

Moreover, this network of seven municipalities has real potential as a cultural slow-tourism destination. The area is already familiar with slow philosophy (e.g. the headquarters of Slow Food is located in Bra) and the preservation of cultural heritage.

Through promoting the symbolic associations that characterize the network, every single node of the network (i.e. each sister-city) can be enriched by new values and perceptions released by the network itself and by other nodes (Pasquinelli, 2017). The network of the Seven Sisters could, therefore, become an interesting slow-tourism alternative to more renowned destinations.

Nevertheless, networking requires a strong commitment and a vision shared by all the stakeholders involved. Usually, stakeholders (especially policy-makers) adopt a collaborative approach when it represents a real added value and possibility for growth, dealing with issues that cannot be addressed within pre-existing administrative-based borders (Boisen, Terlouw & van Gorp, 2011).

Considering the historical and cultural heritage shared by the sister-cities and their potential as a slow-tourism destination, this study selected the Seven Sisters to deepen the issue of networking in slow-tourism development. The Seven Sisters nonstandard region is in a latent stage, and this makes this case a unique opportunity to explore what can facilitate and/or hinder the blossoming of an effective network. The main topics discussed are:

1. Issues and objectives for cooperation among the sister-cities;
2. Potential advantages of a shared network perspective;
3. Obstacles to cooperation in the creation of the Seven Sisters network.

3.3 Research Design, Data Collection, and Analysis

The study adopted a qualitative research methodology because the interest was on deeply exploring a particular case, not still object of analysis (Ritchie, Lewis, Nicholls & Ormston, 2003), related to network-building in the humanistic tourism perspective and local, sustainable development.

In-depth, semi-structured interviews with local stakeholders were conducted face-to-face during April and May 2019, in separate sessions lasting 45 minutes on average. The interviews were digitally recorded. In advance of the interviews, participants were informed about the privacy policies and the confidentiality of their interviews.

Eighteen stakeholders were contacted, with a response rate of 72%. They were selected using theoretical sampling (based on the theoretical needs of the study) and snowball sampling (research participants recommended other potential participants) (Patton, 2002). Being interested in understanding the advantages and obstacles to cooperation in slow-tourism development, the respondents were recruited among actors with political and managerial decision power.

Key informants were divided into institutional and non-institutional actors (Table 2.1). The institutional informants were members of the city councils of each of the sister-cities, while the non-institutional actors were members of important local cultural and socio-economic associations, foundations, and organizations, involved in the territorial development. The division between institutional and non-institutional actors was considered useful in examining different sensitivities to topics such as sustainability, slow tourism, and networking. Each sister-city was represented by at least one institutional actor and one non-institutional actor.

Although the bottom-up approach is a viable option for implementing the slow philosophy in tourism development, private tourist operators and the local community were not considered because network-building and cooperation among municipalities requires as a preliminary condition the commitment of the political and institutional representatives (Boisen, 2015). At the same time, the non-institutional actors were considered because of the supportive and influencing role they can play in the network project.

Table 2.1 Institutional and non-institutional stakeholders interviewed

Institutional stakeholders	
Institution (municipality)	*Role/function*
Cuneo	Municipal councilor
Alba	Municipal councilor
Bra	Municipal councilor
Mondovì	Municipal councilor
Saluzzo	Municipal councilor
Savigliano	Mayor
Fossano	Mayor and municipal councilor
Non-institutional stakeholders	
Association/foundation/organization	*Role/function*
ARTEA Foundation	Director
ATL del Cuneese (Local Tourist Board)	President
Cassa di Risparmio of Cuneo Foundation	President
Cassa di Risparmio of Saluzzo Foundation	President
CATAC Association and Compral Latte Cooperative	President
FAI Fondo Ambiente Italiano—Manta Castle	Property Manager

Source: Author's own elaboration

The interviews were transcribed with a view to conducting two text analyses: the first entailed thematic analysis of data with a systematic read and reread to sum up relevant themes (Boyatzis, 1998; Fereday & Muir-Cochrane, 2006) and a subsequent rationalization of key elements; the second was conducted with T-Lab software, employing statistical and lexical methods to highlight the relevance and frequency of words and topics.

The use of T-Lab software helped to highlight the frequency of the lemmas[1] used, the keywords, and associations between them. Some grouping with an internal homogeneity and an external heterogeneity became necessary to avoid fragmentation resulting from the use of synonyms (Lancia, 2012).

The most relevant keywords related to the topics analyzed in each question are represented in radial diagrams showing, for each lemma, the grade of association with the others considered significant by the software applying a chi-square test ($p < 0.05$). The distance between keywords and other lemmas is proportional to the grade of association: the shorter the distance, the greater the correlation.

4. Findings

The analysis is divided into three sections that deal with themes, potentialities, and obstacles related to cooperation among the sister-cities.

4.1 Issues and Objectives for Cooperation Among the Sister-Cities

With reference to potential cooperation among the Seven Sisters, the majority of actors, both institutional and non-institutional, stressed the same principal topics as a focus for their policies and efforts: the outdoors, gastronomy, events, and culture. At first glance, the presence of the three pillars of tourism attraction emerges: nature, food, and culture. These elements are united by the concept of cooperation, and they are implemented according to slow tourism and local identity perspectives.

It is interesting to note that the majority of actors interviewed consider nature and landscape conservation as the main themes of a shared development among the Seven Sisters and the principal area where slow-tourism policies can be jointly implemented. The following comments exemplify this:

> Relating to slow tourism we go after all the aspects of the outdoor because it is a part of our territory.
>
> (R3)

> About slow tourism there is a real chance to work together with the upcoming project on bike routes of the outdoor tourism, where we are trying to work on the territory.
>
> (R6)

Looking at the radial diagrams of the keyword 'Slow Tourism' there is a confirmation of that association (Figure 2.1). The outdoor activities are always connected, both for institutional and non-institutional actors, to the words 'Slow Tourism'. Other interesting co-occurrences emerging from slow tourism are 'important' and 'potentiality', highlighting the centrality of considering economic and socio-cultural growth only if it involves a sustainable base and a humanistic perspective.

The diagram results, crossed with the text analysis, reveal that slow tourism led by outdoor aspects tend to be more important for institutional actors than non-institutional ones. The latter tend to have a broader vision rather than a specific target related to slow-tourism perspectives. The centrality of the outdoors for policy-makers probably reflects the fact that interviewees had in mind the financial support given by the European Union and other institutions to improve cycle paths and hiking trails, or other investments in sport and nature, rather than supporting culture and tradition.

The priority given to the outdoors by institutional actors is confirmed by the 'Issues' radial diagrams (Figure 2.2). Policy-makers associate 'issues' with 'nature', 'beauty', 'mountains', and 'water', while non-institutional actors tend to be more concerned about the seemingly

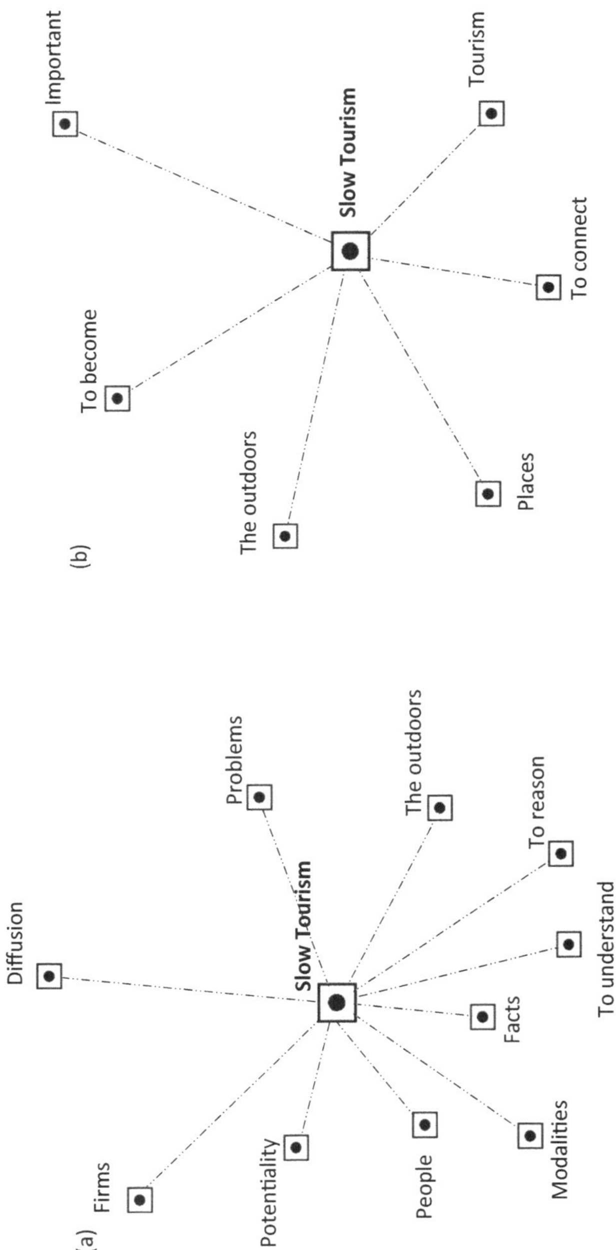

Figure 2.1 Slow tourism: radial diagrams for institutional (a) and non-institutional (b) actors

Source: Author's own elaboration

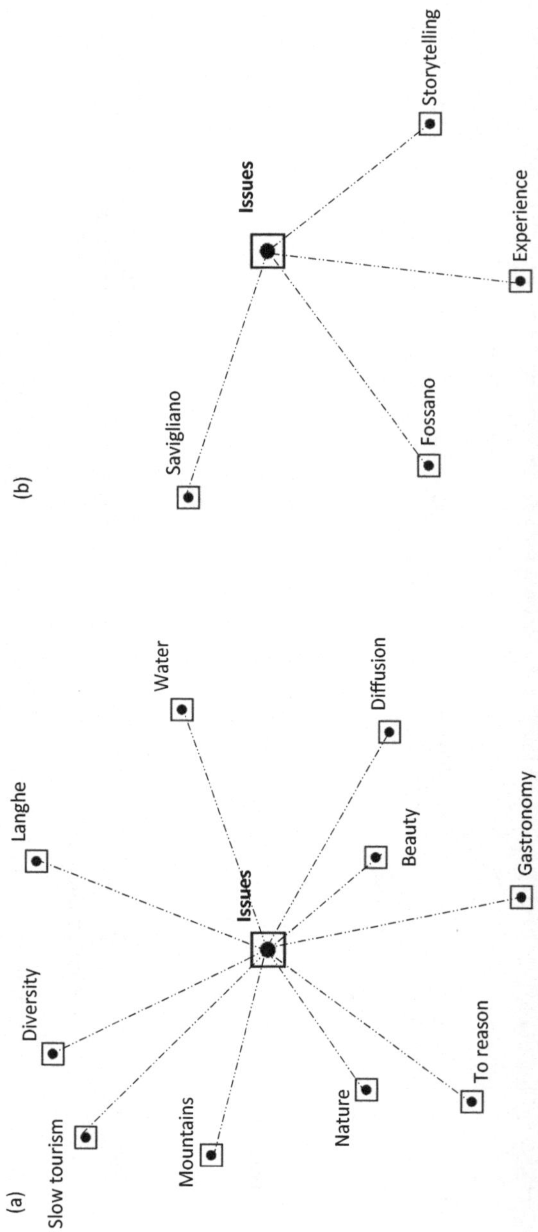

Figure 2.2 Issues: radial diagrams for institutional (a) and non-institutional (b) actors

Source: Author's own elaboration

less distinctive lowland sister-cities of Savigliano and Fossano. Non-institutional stakeholders talk about 'storytelling' and 'experience', highlighting the importance of communication messages targeting the new frontier of experience in tourism management.

The 'Identity' radial diagrams show the institutional actors' focused view of the Seven Sisters' identity, compared with the more holistic view of the non-institutional actors (Figure 2.3). The former underlines the importance of gastronomy and agriculture as identity elements. For non-institutional actors, 'Identity' is more strongly related to other words such as 'diversity', 'cultural', 'citizens', 'Saluzzo', 'find', and 'objective'. A search for convergence among the lemmas shows that 'citizens' are the basic social elements of the 'identity' of the Seven Sisters. This 'identity' becomes an 'objective' and is based on the 'cultural' aspects and the 'diversity' of resources, with a specific link to 'Saluzzo' as the cultural heart of the province of Cuneo.

4.2 Potential Advantages of a Shared Network Perspective

With reference to the potential advantages of cooperation and networking, text analysis results converge toward branding and communication as key elements that can advance the progress of the Seven Sisters. The majority of actors sees the benefits of working together under the 'Seven Sisters' brand, considering place branding helpful for a stronger distinctive image and more efficient communication. The creation of a distinct presence of the 'Seven Sisters' brand in the hearts and minds of people could attract new slow tourism. This would be welcomed by local inhabitants and could attract new citizens. One of the non-institutional actors clearly quoted:

> If the sister-cities work on simplify what they want to communicate, something meaningful and significant for the local identity of each Sister, this will help the creation of a place brand hanging together seven 'characteristics' completing each other.
>
> (R9)

The system of statistically significant associations with the lemma 'Advantages' confirms and supports the text analysis (Figure 2.4). For city managers, 'identity', 'coordination', and 'costs' are co-occurrences, while for non-institutional actors the main words are 'unique concept', 'communication', 'collaboration', 'beauty', 'levels', and 'gastronomy'. This highlights the possibility to 'communicate' a 'unique concept' through 'collaboration' among the sister-cities, based on the 'beauty' of the places and 'gastronomy'. Policy-makers are more concerned about costs; consequently, they see coordination as a cost-sharing opportunity.

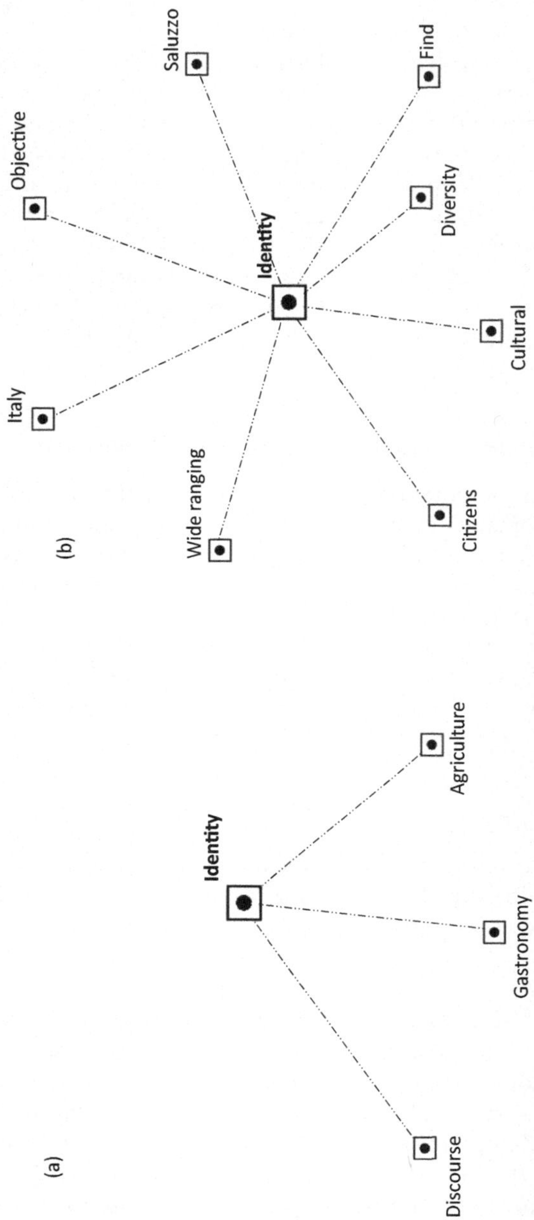

Figure 2.3 Identity: radial diagrams for institutional (a) and non-institutional (b) actors

Source: Author's own elaboration

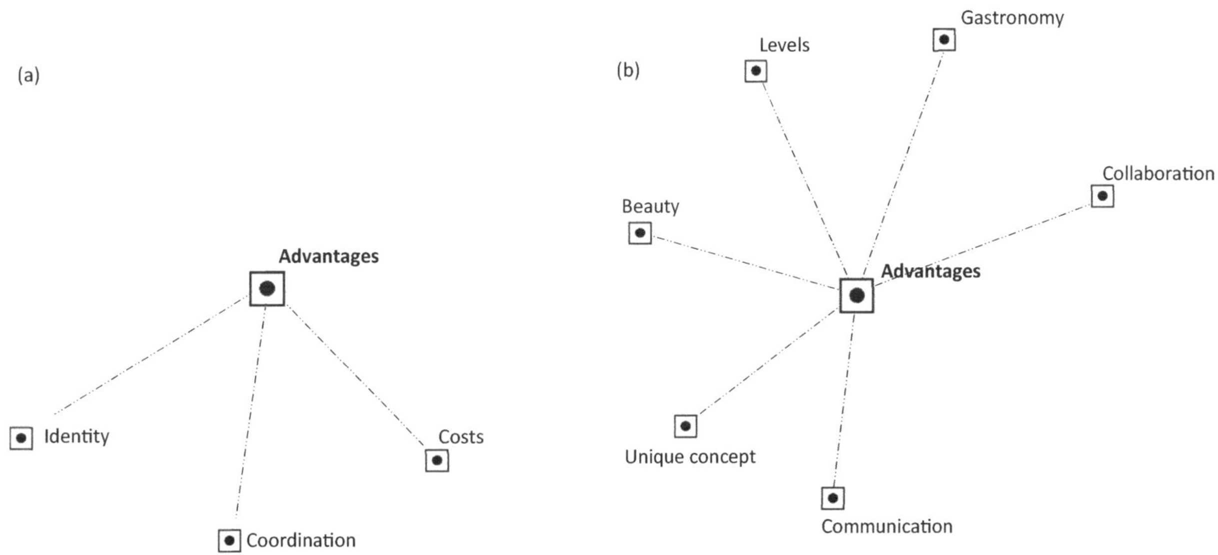

Figure 2.4 Advantages: radial diagrams for institutional (a) and non-institutional (b) actors

Source: Author's own elaboration

4.3 Obstacles to Cooperation in the Creation of the Seven Sisters Network

A convergence around three main problematic issues in creating a sustainable network among the seven municipalities emerges from the interviews.

The first is parochialism, meaning mental closure and the tendency to see only very local questions about one's own city. The main fear is: 'Why should I help the others to succeed, and what is in it for me?' The problem is a difficulty in seeing the big picture, in working toward cooperation with one's neighboring cities. In particular, there are two competing attractions: the mountains (near Mondovì and Cuneo) and the Langhe hills (near Alba). An institutional respondent commented:

> There's a really strong parochialism among us, the Seven Sisters. We do not feel it among us [intending Cuneo and Mondovì], but between us and Alba. Citizens of Alba are in their own way.
>
> (R8)

Indeed, the Langhe territory seems to converge more toward collaborations with other nearby areas such as Monferrato (having common environments and culture) and Turin (one of the strongest cultural poles in Italy).

When there are two strong magnetic poles with opposite energies, it can seem difficult to foresee any kind of collaboration. This makes it difficult to see that the area's attractions could complement one another in a broader mosaic encompassing a worthwhile, sustainable horizon.

The second issue is local politics. Some of the sister-cities are in the hands of opposing political parties, and this can cause rivalry and mistrust. Different opinions and approaches to the same topic can make it difficult to build a shared vision of the future and a long-term perspective on development.

The last obstacle is logistics and infrastructure. In fact, the province of Cuneo suffers from a lack of motorways linking the eastern and western parts of the region. Railways are in need of modernization, and the tunnel connecting the area with France is blocked because of subsidence. All the stakeholders agree on this point, and the following question arises: 'Why cooperate if the infrastructure cannot support tourism?' Both institutional and non-institutional actors do not seem to understand that cooperation is the starting point to overcoming these deficiencies.

As shown in Figure 2.5, institutional actors associate the words 'parochialism', 'infrastructures', and 'rivalries' with the lemma 'Problems', confirming what emerged from critical analysis. Institutional actors also point out the words 'not easy to reach', highlighting logistics and infrastructure as major local problems. Non-institutional actors link

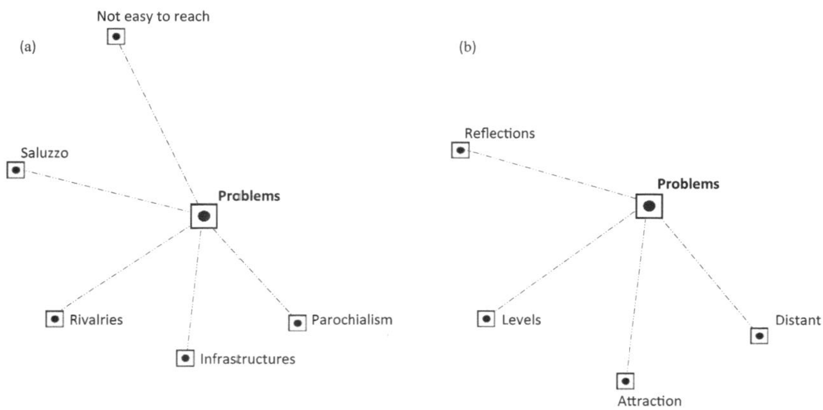

Figure 2.5 Problems: radial diagrams for institutional (a) and non-institutional (b) actors

Source: Author's own elaboration

'Problems' with 'attraction' of customers and related revenue and investments, and with the word 'distant', which is again connected to the infrastructural issues. These types of problems call for 'reflections' at different 'levels', underlining the multi-scalar nature of place governance.

5. Conclusion

The chapter discusses the potential for local tourism development based on a slow positioning and networking among small-scale cities. The potential nonstandard region of the Seven Sisters has been analyzed through key informant interviews. It has been highlighted that sustainable forms of tourism, such as slow tourism, fit with the humanistic management philosophy, having a common body of values. Slow tourism embraces and amplifies people flourishing and well-being, that are core values in a humanistic perspective. Nevertheless, the slow-tourism model seems not easy to be implemented at a network-based scale.

Both institutional and non-institutional interviewees agree that slow tourism makes for common ground among the sister-cities and uniqueness in tourists' perceptions (Heitmann et al., 2011). Nature, landscape, culture, and tradition are seen as areas of potential collaboration and joint policies that could favor sustainable growth and improve overall well-being. Through collaboration, it could be possible to define a unique brand (the 'Seven Sisters') and obtain a distinctive positioning. In addition, cooperation could create cost efficiencies in branding and communication.

In essence, the sister-cities share an historical and cultural common denominator relevant to their internal and external stakeholders, that could strengthen the value proposition of the area, its distinctiveness, and identity (Boisen, 2015). From a humanistic perspective, collaboration among the sister-cities in slow-tourism development is perceived as a valuable way to make the resulting nonstandard region a better place, improving the quality of life for both inhabitants and tourists.

The creation of a network requires strong commitment among participants, who must perceive the relevant advantages of sharing efforts and investments. They must share a common vision of the place and of its future positioning too. Nevertheless, network- and dialogue-based strategies among the sister-cities are limited by parochialism and power fragmentation. The interviewees agree that dialogue and relations are compulsory elements for cooperation, but there is a convergence of opinions that parochialism and individualism are the main obstacle for 'growing together, growing stronger'.

Therefore, the lack of convergent interests and the persistence of parochialism hinder the creation of the Seven Sisters network tied to a humanistic perspective for its goals (i.e. to promote slow tourism as a means to enhance the dignity and well-being of tourists, of people involved in the industry, and of local communities) and its modalities (favoring cooperation vs. competition). The existing literature supports the idea that the most relevant limitation to cooperation are financial issues and loss of political power (e.g. Boisen, 2015; Berg & Lofgren, 2000), but in the Seven Sisters case these factors do not clearly emerge as a major concern.

The research reveals different opinions between institutional and non-institutional actors. Non-institutional actors consider the Seven Sisters as a region with geographical and cultural fluidity, where differences among sub-areas are a strength rather than a limitation. They highlight the three pillars of nature, food, and culture as the basis of the Seven Sisters' value proposition, where the identity of each sister-city is maintained and reinforced as a part of an overall mosaic. Cultural and humanistic aspects, particularly stressed by non-institutional actors, seem to be less important for policy-makers, who tend to focus more on nature-oriented policies.

In conclusion, in the case of the Seven Sisters, two different visions emerge. Non-institutional actors seem ready to act and play a role in a wider context (the network), whereas institutional players are still trying to frame a position and to overcome internal rivalries. The Seven Sisters case confirms that political interest in a network project can be very weak when a cooperative strategy is not motivated by access to public incentives and funding programs, a strong business sector driving local development, and the need for territorial regeneration and requalification (Boisen, 2015).

The chapter addresses a topic so far unexplored. It is subject to limitations that provide avenues for further studies. Future research may involve other categories of stakeholders, such as members of the local business community, citizens, and tourists. Future research can build upon the findings of the present study to investigate the topic in other contexts. As suggested by Ekinci (2014), a comparative study would also be beneficial.

Note

1. Group of words with the same lexical root or belonging to the same grammatical category.

References

Beech, J., & Chadwick, S. (2006). Introduction—The unique evolution of tourism as 'business'. In J. Beech & S. Chadwick (Eds.), *The business of tourism management* (pp. 3–18). Harlow: Pearson Education.

Berg, P. O., & Lofgren, O. (2000). Studying the birth of a transnational region. In P. O. Berg, A. Linde-Laursen, & O. Lofgren (Eds.), *Invoking a transnational metropolis. The making of the Øresund region* (pp. 7–26). Lund: Studentliteratur.

Beritelli, P. (2011). Cooperation among prominent actors in a tourist destination. *Annals of Tourism Research*, *38*, 607–629. doi:10.1016/j.annals.2010.11.015

Boisen, M. (2015). Place branding and nonstandard regionalization in Europe. In S. Zenker & B. Jacobsen (Eds.), *Inter-regional place branding* (pp. 13–23). Cham: Springer. doi:10.1007/978-3-319-15329-2_2

Boisen, M., Terlouw, K., & van Gorp, B. (2011). The selective nature of place branding and the layering of spatial identities. *Journal of Place Management and Development*, *4*, 135–147. doi:10.1108/17538331111153151

Boyatzis, R. E. (1998). *Transforming qualitative information: Thematic analysis and code development*. Thousand Oaks, CA: Sage.

Brenner, N. (2004). *New state spaces: Urban governance and the re-scaling of statehood*. Oxford: Oxford University Press.

Buffa, F., Beritelli, P., & Martini, U. (2018). Project networks and the reputation network in a community destination. Proof of the missing link. *Journal of Destination Marketing & Management*, *11*, 251–259. doi:10.1016/j.jdmm.2018.05.001

Caffyn, A. (2012). Advocating and implementing slow tourism. *Tourism Recreation Research*, *37*, 77–80. doi:10.1080/02508281.2012.11081690

Clancy, M. (2015). *What's at stake in the slow movement? Consumption, responsibility and citizenship*. Chicago, IL: The Association of American Geographers Annual Meeting.

Conway, D., & Timms, B. (2012). Are slow travel and slow tourism misfits, compadres or different genres? *Tourism Recreation Research*, *37*, 71–76. doi:10.1080/02508281.2012.11081689

Cynarski, W. J., & Obodynski, K. (2004). Tourism in humanistic perspective—scientific conference. *Tourism Today, Fall* (4), 170–173.

de la Barre, S. (2012). Travellin' around on Yukon time in Canada's north. In S. Fullagar, K. Markwell, & E. Wilson (Eds.), *Slow tourism: Experiences and mobilities* (pp. 157–169). Bristol: Channel View Publications.

Deas, L., & Lord, A. (2006). From a new regionalism to an unusual regionalism? The emergence of non-standard regional spaces and lessons for the territorial reorganisation of the state. *Urban Studies*, *43*, 1847–1877. doi:10.1080/00420980600838143

Dierksmeier, C. (2011). Reorienting management education: From the homo economicus to human dignity. *Humanistic Management Network, Research Paper Series No. 13–05*. http://dx.doi.org/10.2139/ssrn.1766183

Ekinci, M. B. (2014). The Cittaslow philosophy in the context of sustainable tourism development: The case of Turkey. *Tourism Management*, *41*, 178–189. doi:10.1016/j.tourman.2013.08.013

Fereday, J., & Muir-Cochrane, E. (2006). Demonstrating rigor using thematic analysis: A hybrid approach of inductive and deductive coding and theme development. *International Journal of Qualitative Methods*, *5*, 80–92. doi:10.1177/160940690600500107

Franch, M., & Martini, U. (Eds.). (2013). *Management per la sostenibilità dello sviluppo turistico e la competitività delle destinazioni*. Bologna: Il Mulino.

Fullagar, S., Markwell, K., & Wilson, E. (2012). Starting slow: Thinking through slow mobilities and experiences. In S. Fullagar, K. Markwell, & E. Wilson (Eds.), *Slow tourism: Experiences and mobilities* (pp. 1–8). Bristol: Channel View Publications.

Gibson, D., Pratt, S., & Movono, A. (2012). Tribe tourism: A case study of the Tribewanted project in Vorovoro, Fiji. In S. Fullagar, K. Markwell, & E. Wilson (Eds.), *Slow tourism: Experiences and mobilities* (pp. 185–200). Bristol: Channel View Publications.

Guercini, S., & Tunisini, A. (2017). Regional development policies. In H. Hakansson & I. Snehota (Eds.), *No business is an island* (pp. 141–155). Bingley: Emerald Publishing.

Hatipoglu, B. (2014). "Cittaslow": Quality of life and visitor experiences. *Tourism Planning & Development*, *12*, 1–17. doi:10.1080/21568316.2014.960601

Heitmann, S., Robinson, P., & Povey, G. (2011). Slow food, slow cities and slow tourism. In P. Robinson, S. Heitmann, & P. Dieke (Eds.), *Research themes for tourism* (pp. 114–127). Wallingford: CABI. doi:10.1079/9781845936846.0114

Honoré, C. (2005). *In Praise of slow: How a worldwide movement is challenging the cult of speed*. London: Orion Books.

Howard, C. (2012). Speeding up and slowing down: Pilgrimage and slow travel through time. In S. Fullagar, K. Markwell, & E. Wilson (Eds.), *Slow tourism: Experiences and mobilities* (pp. 11–24). Bristol: Channel View Publications.

Jensen, O. B. (2007). Culture stories: Understanding cultural urban branding. *Planning Theory*, *6*, 211–236. doi:10.1177/1473095207082032

Jepson, A., Stadler, R., & Spencer, N. (2019). Making positive family memories together and improving quality-of-life through thick sociality and bonding at local community festivals and events. *Tourism Management*, *75*, 34–50. doi:10.1016/j.tourman.2019.05.001

Kalandides, A. (2012). Place branding and place identity. An integrated approach. *Tafter Journal*, *43*. Retrieved from www.tafterjournal.it/2012/01/03/place-branding-and-place-identity-an-integrated-approach/

Kavaratzis, M., & Hatch, M. J. (2013). The dynamics of place brands: An identity-based approach to place branding theory. *Marketing Theory*, *13*, 69–86. doi:10.1177/1470593112467268

Lancia, F. (2012). *The logic of the T-LAB tools explained*. Retrieved from www.tlab.it/en/toolsexplained.php.

Lowry, L. L., & Lee, M. (2016). CittaSlow, slow cities, slow food: Searching for a model for the development of slow tourism. *Travel and Tourism Research Association: Advancing Tourism Research Globally*, Paper 40. Retrieved from https://scholarworks.umass.edu/ttra/2011/Visual/40/

Lucarelli, A. (2018). Place branding as urban policy: The (im)political place branding. *Cities*, *80*, 12–21. doi:10.1016/j.cities.2017.08.004

Mayer, H., & Knox, P. L. (2006). Slow cities: Sustainable places in a fast world. *Journal of Urban Affairs*, *28*, 321–334. doi:10.1111/j.1467-9906.2006.00298.x

Melé, D. (2003). The challenge of humanistic management. *Journal of Business Ethics*, *44*(1), 77–88.

Melé, D. (2016). Understanding humanistic management. *Humanistic Management Journal*, *1*, 33–55. doi:10.1007/s41463-016-0011-5

Metzger, J. (2013). Raising the regional leviathan: A relational-materialist conceptualization of regions-in-becoming as publics-in-stabilization. *International Journal of Urban and Regional Research*, *37*, 1368–1395. doi:10.1111/1468-2427.12038

Moore, K. (2012). On the periphery of pleasure: Hedonics, eudaimonics and slow travel. In S. Fullagar, K. Markwell, & E. Wilson (Eds.), *Slow tourism: Experiences and mobilities* (pp. 25–35). Bristol: Channel View Publications.

Murayama, M., & Parker, G. (2012). 'Fast Japan, slow Japan': Shifting to slow tourism as a rural regeneration tool in Japan. In S. Fullagar, K. Markwell, & E. Wilson (Eds.), *Slow tourism: Experiences and mobilities* (pp. 170–184). Bristol: Channel View Publications.

Nilsson, J.H., Svard, A.C., Widarsson, A., & Wirell, T. (2007). *Slow destination marketing in small Italian towns*. Paper presented at the 16th Nordic Symposium in Tourism and Hospitality Research, September 27–29, Helsingborg, Sweden.

Nilsson, J. H., Svard, A. C., Widarsson, A., & Wirell, T. (2011). 'Cittáslow' eco-gastronomic heritage as a tool for destination development. *Current Issues in Tourism*, *14*, 373–386. doi:10.1080/13683500.2010.511709

Oh, H., Assaf, A. G., & Baloglu, S. (2016). Motivations and goals of slow tourism. *Journal of Travel Research*, *55*, 205–219. doi:10.1177/0047287514546228

Park, E., & Kim, S. (2015). The potential of Cittaslow for sustainable tourism development: Enhancing local community's empowerment. *Tourism Planning & Development*, *13*, 351–369. doi:10.1080/21568316.2015.1114015

Pasquinelli, C. (2015). Network brand and branding: A co-opetitive approach to local and regional development. In S. Zenker & B. Jacobsen (Eds.), *Inter-regional place branding* (pp. 39–49). Cham: Springer. doi:10.1007/978-3-319-15329-2_4

Pasquinelli, C. (2017). *Place branding. Percezione, illusione e concretezza*. Roma: Aracne.

Patton, M. Q. (2002). *Qualitative research & evaluation methods*. London: Sage.

Perdue, R. R., Tyrrell, T., & Uysal, M. (2010). Understanding the value of tourism: Conceptual divergence. In D. Pearce & R. Butler (Eds.), *Tourism research: A 20:20 vision* (pp. 123–134). Oxford: Goodfellow Publishers.

Presenza, A., Abbate, T., & Micera, R. (2015). The Cittaslow movement: Opportunities and challenges for the governance of tourism destinations. *Tourism Planning & Development*, *12*, 479–488. doi:10.1080/21568316.2015.1037929

Ritchie, J., Lewis, J., Nicholls, C. M., & Ormston, R. (2003). *Qualitative research practice: A guide for social science students and researchers.* London: Sage.

Santagata, W. (2002). Cultural districts, property rights and sustainable economic growth. *International Journal of Urban and Regional Research*, *26*, 9–23. doi:10.1111/1468-2427.00360

Scheyvens, R. (2002). *Tourism for development: Empowering communities.* Essex: Pearson Education.

Serdane, Z. (2019). Slow philosophy in tourism development in Latvia: The supply side perspective. *Tourism Planning & Development*, 1–18. doi:10.1080/21568316.2019.1650103

Sidali, K. L., & de Obeso, M. (2018). Successful integration of slow and sustainable tourism. In M. Clancy (Eds.), *Slow tourism, food and cities: Pace and the search for the "good life"* (pp. 169–180). Oxon: Routledge.

Spitzeck, H. (2011). An integrated model of humanistic management. *Journal of Business Ethics*, *99*, 51–62. doi:10.1007/s10551-011-0748-6

Spitzeck, H., Amann, W., Pirson, M., Khan, S., & Von Kimakowitz, E. (Eds.). (2009). *Humanism in business.* Cambridge: Cambridge University Press.

Su, L., Huang, S., & Huang, J. (2018). Effects of destination social responsibility and tourism impacts on residents' support for tourism and perceived quality of life. *Journal of Hospitality & Tourism Research*, *42*, 1039–1057. doi:10.1177/1096348016671395

Terlouw, K. (2009). Rescaling regional identities: Communicating thick and thin regional identities. *Studies in Ethnicity and Nationalism*, *9*, 452–464. doi:10.1111/j.1754-9469.2009.01064.x

Timms, B. F., & Conway, D. (2012). Slow tourism at the Caribbean's geographical margins. *Tourism Geographies*, *14*, 396–418. doi:10.1080/14616688.2011.610112

Uysal, M., Sirgy, M. J., Woo, E., & Kim, H. L. (2016). Quality of life (QOL) and well-being research in tourism. *Tourism Management*, *53*, 244–261. doi:10.1016/j.tourman.2015.07.013

3 Do Socially Responsible Tour Operators Improve Tourism Sustainability?

The Case of Nairobi, Kenya

Simon M. Thiong'o, Buke Bashuna, Bob E.L. Wishitemi, and Nehemiah Kiprutto

1. Introduction

According to the United Nations and the World Trade Organization, tourism is a very dynamic industry, contributing 5% of the world GDP and 6–7% of the working places only from direct activities. Taking into account the indirect and induced effects generated by the strong multiplier character of the tourist industry, the GDP contribution is 9%. The great contribution made by the tourism industry has necessitated that its stakeholders be keen in their operations so as to maintain and boost further the benefits associated by efficient tourism.

Tour operators play a leading role in the travel and tourism industry by developing tourism products and offering them to tourists (Budeanu, 2005). Over the past decades, many tour operators have incorporated operational strategies in operations that have in turn contributed to sustainable tourism development. According to Holloway (2010), tour operators are tagged with the responsibility of providing convenient package options for tourists, and to combine these components into a tour. Tourism sustainability has, nonetheless, posed a great problem in the industry and has also contributed to the little progress experienced today.

Buhalis (2001), however, notes that effective role-play by tour operators as one of the stakeholders in the tourism industry can create sustainable tourism development and in the long run help to curb the problem of unsustainability. Globally, many theories (corporate public policy, business ethics, and stakeholder theory/management, humanistic) have been put forward relating to the roles played by tour operators in sustainable tourism development. The theories seem to suggest that these roles do not enhance sustainable tourism development, while others have suggested that tour operators' functions are best positioned to enhance sustainable tourism (Mitchell & Ashley, 2010). A proper balance of the roles played by tour operators is a more appropriate way to curb the issues of unsustainability in tourism development because it offers a means of solving the existing problem of sustainable tourism.

Dahiru (2011) noted that in Africa, the poor performance of tour operators has been linked to tourism sustainability-related challenges. For example, the emergence of an increasingly sophisticated market niche—more experienced, well-educated and aware of what competition has to offer ("children of information age")—has brought about the challenge of changing travel patterns from destination-oriented to experience-oriented in South Africa. This has affected the general performance of South African tour operators and the tourism industry, despite the fact that the country possesses incredible tourist attraction sites and good infrastructural facilities.

Regionally, existing literature has not been able to determine the influence of tour operators' activities on sustainable tourism development. Locally, the creation of sustainable tourism through efficient role-play by tour operators has not borne much fruit as many tour operators are negligent and lag behind in carrying out their roles and effect on sustainable tourism development. Omollo (2012) has noted that attainment of sustainable tourism development by many tour operators in Nairobi is attributed to the professional management of the operators, but Dent (2012) has argued that the creation of sustainable tourism development is dependent entirely on the roles of key stakeholders, especially the ministry of tourism, in curbing the problem of unsustainability in the industry. Most tour-operating companies in Kenya have relied heavily on seasonal peaks to grow, which has proven fatal to business because of a lack of stability. This, therefore, implies the need to undertake a carefully considered mode of assessing the role of tour operators and their effect on sustainable tourism development despite perceived perceptions associated with its seasonal nature. For instance, in Kenya, tour operators such as Vintage Africa Ltd, among others, work with local communities to manage wastes and emissions (Akama, 1997). This has led to conservation of tourist attraction sites which has greatly contributed to sustainable tourism development.

Generally, tourism has been used by many countries as a panacea to development (Cheng & Zhang, 2020; Thiong'o, 2019; Maroto-Martos, Voth & Pinos-Navarrete, 2020). However, decades of developing tourism has led to the realization that it is a double-edged sword—containing seeds of its own destruction (Ahmad, Draz, Su & Rauf, 2019; Gössling, 2003; Mishra & Kestwal, 2019). Proper management and planning is, therefore, required to maximize its benefits and minimize its negative impacts (Mayne, 1997). Given the importance of tourism in local and global economies, there is need to ensure that the base resources for tourism and the benefits accrued are sustained (Ristić, Vukoičić & Milinčić, 2019; Ribot, 2004).

Most tour operators point a blaming finger on the failure of tourism stakeholders to promote sustainable tourism as they attempt to curb problems associated with managing the tourism industry (Kwena, 1997). Proven facts are lacking as a way to improve the tourism industry. It can be argued that the unavailability of such information potentially deprives operators of the information they need to curb this problem. Omollo (2012) notes that inefficiency by the tour operators prevents the tourism industry from creating an enabling environment for further developments.

Tour operators play a leading role in the travel and tourism industry by developing tourism products and offering them to tourists (Wall-Reinius, Ioannides & Zampoukos, 2019; Drobotova, Krasnomovets, Radchenko & Romanov, 2019). Thus, they should be at the forefront in ensuring best practice in destinations. The role played by tour operators to curb unsustainable tourism has, however, been insufficient and at best, the existing literature does not show the link between tour operators' roles (particularly corporate social responsibility [CSR]) and sustainable tourism development. This deficiency prompted this study.

2. Literature Review

Sustainable tourism is the concept of visiting a place as a tourist and trying to positively affect the environment, society, and economy (Lisse, 2010). Tourism can involve primary transportation to a general location, accommodation, entertainment, recreation, and shopping, among others. It can be related to travel for leisure, gastronomy, business, and visiting friends and relatives (VFR) (Peeters & Dubois, 2010). Thus, there is a broad consensus that tourism development should be sustainable; however, the question of how to achieve this remains an object of debate (Peeters et al., 2004).

Without travel, there is no tourism. The concept of sustainable tourism development, therefore, is tightly linked to a concept of sustainable mobility (Høyer, 2000). Two relevant considerations are tourism's reliance on fossil fuels and tourism's effect on climate change—72% of tourism's CO_2 emissions come from transportation, 24% from accommodation, and 4% from local activities (Peeters & Dubois, 2010). However, when considering the impact of all greenhouse gas emissions from tourism and that aviation emissions occur at high altitudes where their effect on climate is amplified, aviation alone accounts for 75% of tourism's climate impact (Gossling, Hall, Peeters and Scott, 2010).

The International Air Transport Association (IATA) considers an annual increase in aviation fuel efficiency of 2% per year through 2050 to be realistic. By 2050, with other economic sectors having greatly reduced their CO_2 emissions, tourism is likely to generate 40% of global carbon emissions (Cohen, Higham, Peters and Gossling, 2014). The main cause is an increase in the average distance traveled by tourists, which for many years has been increasing at a faster rate than the number of trips taken.

Sustainable transportation is now established as the critical issue confronting a global tourism industry that is palpably unsustainable (Gossling et al., 2010).

Sustainable tourism development is envisaged as leading to management of all resources in such a way that economic, social, and aesthetic needs can be fulfilled while maintaining cultural integrity, essential ecological processes, and biological diversity and life support systems (McIntyre, 1993). Sustainable tourism products are those that are operated in harmony with the local environment, communities, and cultures so that these become the beneficiaries and not the victims of tourism development (Høyer, 2000).

The Prosper model developed by the Sustainable Tourism Cooperative Research Centre (STCRC) uses an indicators approach to assess the economic, social, and environmental value of tourism in a destination. The STCRC kit for Promoting Awareness of the Value of Tourism is an important tool for communicating with stakeholders. The components include: industrial contribution—business investment; social contribution—community participation, civic pride; municipal contribution—infrastructure management, urban planning; cultural contribution—maintenance of regional image, heritage, and cultural resources; capacity contribution—partnership establishment, data collection, cooperative ventures; environmental contribution—preservation of natural environments; and tourist contribution—visitor numbers and satisfaction.

2.1 Sustainability Theory and Its Humanistic Perspectives

This study was guided by the sustainability theory which employed the use of three models: the economic model, the ecological model, and the political model (Griessler & Littig, 2005; UNDESA, 2007a, 2007b). The theory tends to prioritize and integrate social responses to environmental and cultural problems (Barry, 2007). The economic model looks to sustain natural and financial capital; the ecological model looks into biological diversity and ecological integrity; and the political model looks into social systems that realize human dignity.

The economic model as developed by Solow (1993) proposes that sustainability ought to be thought of as an investment problem, in which we must use returns from the use of natural resources to create new opportunities of equal or greater value. From Solow's classical viewpoint, social spending on the poor or on environmental protection, while perhaps justifiable on other grounds, takes away from this investment and so competes with a commitment to sustainability (Bauman, Bohannon & O'Brien, 2017). With another view of capital, however, the economic model might look different. If we do not assume that "natural capital" is always interchangeable with financial capital, and other proponents of ecological economics, argues Herman (1996), then sustaining opportunity for the future requires strong conservation measures to preserve ecological goods and to keep economies operating within natural limits. These considerations complement an ecological model. From a different perspective of the relation between opportunity and capital, spending on the poor might be regarded as a kind of investment in the future. According to Amartya (1999), we create options for the future by creating options for today's poor because more options will drive greater development.

The ecological model by Olson (1994) proposes the sustainment of ecological integrity and biological diversity. That is, rather than focusing on opportunity as the key unit of sustainability, they focus directly on the well-being of the world. From an anthropocentric viewpoint, the pertinent naturally occurring resources, along with regenerative processes and ecological systems on which human systems depend, should be supported and especially, sustained (Bauman et al., 2017; Tøllefsen, 2011). Further, and according to Tøllefsen (2011), species and ecological systems ought to be sustained for their intrinsic value, and as generators of creatures with intrinsic value, respectively (eco-centrism).

The political model as asserted by Agyeman (2005) proposes sustainment of social systems that realize human dignity. Additionally, there is a focus on sustaining the environmental conditions for a sustainable human life—in which local and global environmental problems jeopardize human dignity (UNECOSOC, 2008; Scoones, 2016). Essentially, this model focuses on and advocates for environmental and civic justice, although the model's pragmatic approach in regard to keeping the sustainability debate alive is constrained by the fact that ecological and economic quality and quantity is regulated by the political system in play (Jenkins, 2009).

Summarily, in addressing tourism sustainability, decision-makers and tourism stakeholders need to constantly be mindful of the complementarities and relationships among these three distinct yet interconnected pillars. It is also necessary that all players are sustainable-development conscious, cultured, and compliant. This is why Hoffman (2000) asserts, "Sustainable development will not become a genuine concern until the business environment becomes a driver of the social equity issues inherent in the sustainability agenda."

In relation to this study, the theory, therefore, can be used to show that sustainability is an investment problem. For tour operators to generate "supportable tourism", they should prioritize and integrate social responses to environmental, economic, and cultural (and political) problems. They should also focus on effective role-play so as to promote the good and positive aimed at harnessing sustainable tourism development. Moreover, while tourism has positive implications for societies, the industry raises many concerns. Given this, actors in the industry should embrace a "slow tourism" configuration (Valls, Mota, Vieira &

Santos, 2019)—that which is capable of enhancing responsible practice on social-cultural environments within which they operate (Miretpastor, Peiró-Signes, Segarra-Oña & Mondéjar-Jiménez, 2015). This is based on the understanding by Gray, Kouhy and Lavers (1995) that the environment has a limited natural capital. Thus, an environmentally (and therefore socially) sustainable society protects and ensures that the systems are kept running for the better good of everyone. This is particularly so in developing countries (Kenya in this case) where the relative importance and reliance on natural resources is greater (Cistulli, 2002)—an implication that the relationship between the environmental resources base and welfare is pertinent.

Moreover, tourism enterprises (tour operators in this instance) influence the environment within which they operate (Niedziółka, 2014). There is, therefore, a general expectation that they should not only influence economic growth but also generally contribute to the welfare of societies (social responsiveness)—directly or indirectly (Kornienko, 2017). Essentially, there should be a mutual trade-off between tour operators and the society through doing what is generally acceptable and morally right—that is, humanism (Caton, 2016).

There could, therefore, be a universal understanding that "good" practice for tour operators (and other tourism stakeholders) ought to be anchored in an established and reflective approach that represents the actual nature of giving back—a humanistic, common-sense approach—that is, social responsibility (DeCarvalho, 1991; Caton, 2016; Goldstein, 1986). This could ultimately ensure responsible human behavior (Mensah & Ricart Casadevall, 2019).

2.2 *Establishing the Place of Corporate Social Responsibility in Sustainable Tourism Development*

Corporate social responsibility (CSR)—used here to refer to the contribution towards societal goals and practices—is a valuable tool for creating reputational capital (Fisher, 2004). Tour operators accepting CSR means acknowledging that they have a responsibility not only to their owners and shareholders but also to society (Kornienko, 2017).

According to studies such as Font and Cochrane (2005) and Khairat and Maher (2012), among others, tour operators have in the past, tended to neglect their socio-environmental responsibilities, citing as justification that they are mere intermediaries, and that destination impacts are a local authority responsibility. While these stakeholders clearly share the responsibility, most tour operators now understand that it is precisely because they are intermediaries (working closely with tourists and tourism service suppliers) that they can have an important impact on management of destination resources by influencing consumer choices, suppliers, and the development patterns of destinations (Curtin & Busby, 1999;

Styles, SchÖnberger & Galvez Martos, 2013; Dent, 2012). Tour operators, thus, need to incorporate CSR in their operations at all times so as to boost industry sustainability. Besides, their practice of CSR equates to a good corporate image and reputation (Foss, 1997).

Rogovysky and Dunfee (2002) posit that the tourism business, just like any other business, will be spurred to adopt CSR in order to fulfill a strategic-moral need. Lavelle (2002) agrees with this and adds that CSR can be used by tour operators as a strategic tool if the program is aligned to the tour operator's goal, resulting in a competitive edge over competitors (corporate social performance) in the same way that price differentials can be used (Rangan, Chase & Karim, 2012). He further argues that where CSR is successfully woven into tour operators' goals, the effect of enhancement on the brand name cannot be challenged. For Picket (2008), this is a smarter way of gaining mileage from public relations because it results in a win-win situation, with the tour-operating company being viewed more favorably by both society at large and shareholders.

Viewpoint (2008) further argues that the management of tour-operating firms can be forced to adopt CSR to remain relevant. This is fulfilling the moral goal, by doing what is right in the eyes of society ("humanism"). Doing what is right puts a burden of adoption of CSR on tour operators because they are perceived as taking from society and should, therefore, give back to society. Additionally, they may be better suited to deal with numerous societal problems than governments or non-governmental organizations as they operate in competitive markets and develop unique competencies (Fernando, 2007). They can also have a large local-community knowledge base, which may not be matched by governments or non-governmental organizations. Smith (2007) limits this capability mainly to large tour-operating firms, as small tour-operating firms tend to have limited resources and, hence, limited capacity.

In reference to the theory, and in addressing the study question, the following conditions were assumed to be true for the study to be undertaken:

i. That the tour operators are in a position to enhance sustainable tourism development by the roles they play in the tourism industry.
ii. That these tour operators have in the past encountered the challenge of unsustainable tourism and are aware of its impact on their business, hence are ready to employ sustainable tourism development strategies to curb this.
iii. That the tour operators value tourism and are keen to promote it so as to curb unsustainable tourism.

The *hypothesis* for the study was, therefore: "There is no significant relationship between Tour operators' practice of Corporate Social Responsibility and sustainable tourism development in Nairobi County."

3. Methods

3.1 Research Design

The study adopted a survey research design because of its ability to collect varied responses from the respondents, with an aim of properly understanding the issues under study. This implies that, through the survey study, the researchers were able to examine in detail the contribution of tour operators and their effect on sustainable tourism development.

Survey study research excels at bringing understanding of a complex issue or object and can extend experience or add strength to what is already known through previous research (Patton, 2002). According to Kerlinger (1973), descriptive studies are not only restricted to fact finding but may also be used in the formulation of important principles of knowledge and, hence, the applicability of survey method.

The target population consisted of 1,440 employees of tour-operator firms and the Kenya Association of Tour Operators representatives (Table 3.1).

3.2 Sampling Procedures

The study adopted a systemic sampling method to draw a sample of the desired number of tour operators to include in the study. All tour operators in Nairobi County were computer generated in alphabetical order, for a total of 60. The third tour operator from the existing list was picked systematically, hence drawing 20 tour operator firms in Nairobi:

N^{th} term = 3 $\quad$ 60/3 = 20

The study employed simple random sampling for the employees and a purposive sampling technique for KATO employees and managers. This technique enabled the researchers to sample one KATO representative, four managers of tour-operator firms, and 298 employees of tour-operator firms.

Table 3.1 Target population

Category	*Number of employees*
KATO employees	5
Managers of KATO[1]-registered tour firms	20
Employees of KATO-registered tour firms	1,415
Total	**1,440**

Source: Tourism Regulatory Authority (2016)

The sample size of the study was calculated using the following formula as recommended by Fisher (1998) and Fan and Thompson (2001):

$$nf = \frac{n}{1 + \frac{n}{N}}$$

Where: nf = Sample size (when the population is less than 10,000).
n = Sample size (when the population is more than 10,000); 384 (a constant)
N = Estimate of the population size; 85

3.3 Sample Size for the Respondents

The desired sample size, therefore, comprised 303 respondents (Table 3.2).

3.4 Data Collection Procedures

The researchers obtained a research permit from Kenya's National Commission for Science, Technology and Innovation (NACOSTI) before proceeding to data collection. After participation in the study was confirmed, a date was set and appointment booked with the organization authorities as well as the study participants. Questionnaires were administered to the employees of the tour-operating firms, utilizing a five-point Likert scale—Strongly Agree (SA), Agree (A), Neutral (N), Disagree (D), and Strongly Disagree (SD)—which was assigned scores of between 5 and 1, respectively. This allowed the researchers to draw conclusions based on comparisons made from the responses.

The researchers then conducted face-to-face interviews with the managers of the tour firms and KATO representatives. Interviews eliminated any sources of bias that could be associated with the use of questionnaires—there were opportunities to clarify any misunderstandings through

Table 3.2 Sample frame of the respondents

Category	*Target population*	*Procedures*	*Sample frame*	*Sampling procedures*
KATO representatives	5	5/1,440 × 303	1	Purposive sampling
Managers of tour firms	20	20/1,440 × 303	4	Purposive sampling
Employees	1,415	1,415/1,440 × 303	298	Simple random sampling
Total	**1,440**		**303**	

Source: Authors' own elaboration

probing (respondents also had freedom in answering). The assembled information was collected for analysis.

3.5 Description of Data Analysis Procedure

Qualitative analysis was done thematically to analyze data collected for interview schedules. The strength of qualitative research is its ability to provide complex textual descriptions of how people experience a given research issue (Mugenda & Mugenda, 2003; Denzin, 2000). It is also effective in identifying intangible factors, such as social norms, socio-economic status, gender roles, ethnicity and religion, whose role in the research issue may not be readily apparent (Denzin, 2000).

Quantitative techniques were done using descriptive and inferential statistics. Descriptive statistics involved the use of frequencies, percentages, and means, while inferential statistics involved the use of the following regression model:

$$Y = \beta_0 + \beta_1 X_1 + \beta_2 X_2 + \beta_3 X_3 + \beta_4 X_4 + e$$

X1: Corporate social responsibility
X2: Community involvement
X3: Participating in responsible tourism
X4: Participation in developing tourism policies
β0: Constant term
β: Beta coefficients
e: Error term

4. Data Analysis, Presentation, and Interpretation

4.1 Overview

Targeting 1,440 respondents, the study sought to investigate CSR as the responsibility of tour operators and its effect on sustainable tourism in Nairobi, Kenya. From the possible sample size of 303, the study managed to collect date from 287 respondents. This translated to a 94.7% response rate, which was considered sufficient for data analysis.

4.2 Background Information of the Respondents

Understanding study respondents' social and economic demographics is critical in conceptualizing the nature of, and characteristics of respondents, which may influence results. Through analysis frequency and percentages, the study identified the background information of the respondents, including age bracket and level of experience in the industry. The findings are displayed in Table 3.3.

Table 3.3 **Background information of the respondents**

Gender	*Frequency*	*Percent*
Age		
Less than 30 years	61	21.3
31–40 years	81	28.2
41–50 years	92	32.1
More than 51 years	53	18.5
Total	287	100.0
Length of experience		
Less than 3 years	53	18.5
4–7 years	107	37.3
8–11 years	97	33.8
More than 11 years	30	10.5
Total	**287**	**100.0**

Source: Authors' own elaboration

The findings regarding age bracket of the respondents showed that 21.3% were younger than 30 years, 28.2% were between 31–40 years, 32.1% were between 41–50 years, and 18.5% were older than 51 years. This implies that the researchers were able to avoid bias in terms of age group by collecting data across different age groups and representing the opinions of valid age groups.

Results regarding length of experience of the respondents indicated that 18.5% had worked fewer than three years, 37.3% between four and seven years, 33.8% between eight and 11 years, and 10.5% had worked for more than 12 years in the industry. This implies that the majority of the respondents had served between four and seven years, indicating that employee retention was high and turnover minimal, thereby maintaining a competent and experienced human resource base—an important aspect in this study. The study collected varied opinions, and the responses were deemed a true (or at least fair) representation of the happenings at the institutions, without influences resulting from being with an institution for too long or being relatively new to the organization.

4.3 Analysis of the Study Objective

4.3.1 Effect of Corporate Social Responsibility on Enhancing Sustainable Tourism

The study first sought to determine the reliability of the variables of corporate social responsibility; the results are given in Table 3.4.

Reliability results indicated that the variables had a Cronbach's coefficient alpha value of 0.805 which was above 0.70; therefore, the study variables were valid for analysis.

Table 3.4 Reliability results of corporate social responsibility

Reliability statistics	
Cronbach's alpha	*Number of items*
0.805	4

Source: Authors' own elaboration

Table 3.5 Effect of corporate social responsibility on sustainable tourism

Corporate social responsibility		*Total*	
Ensures tour operators participate in the rehabilitation of tourist attraction sites	Freq	287	3.47
	%	100	69.4
Enables tour operators to fund tourism-marketing activities	Freq	287	3.44
	%	100	68.8
Ensures tour operators provide quality services in the tourism sector	Freq	287	3.35
	%	100	67
Ensures tour operators reach out to their surrounding society	Freq	287	3.51
	%	100	70.2

Source: Authors' own elaboration

The results (Table 3.5) indicated that 69.4% of the respondents were of the opinion that CSR ensured that tour operators participate in the rehabilitation of tourist attraction sites; 68.8% were of the opinion that SCR enabled tour operators to fund tourism-marketing activities; and 67% were of the opinion that CSR ensured that tour operators provide quality services in the tourism sector.

Further, 70.2% of the respondents were of the opinion that CSR ensured that tour operators reached out to their surrounding society. This implies that the society/environment affects the creation of a sustainable tourism industry.

These findings are in line with the results by Sharma and Talwar (2005) who acknowledge that the tourism business world is dynamic, with tour-operating companies compelled to keep up with changes taking place in the environment. This is further supported by a study by Viewpoint (2008) who presents social responsibility as a critical element to tour-operator activities—that the society would expect them to return benefits accrued from their activities to maintain their relevance.

Results on whether CSR increases participation in rehabilitation of tourist attraction sites (Figure 3.1) indicates that 54% of the respondents agreed with the opinion, whereas 46% disagreed. This indicates that the majority confirms that CSR impacts the increase in participation of the rehabilitation of tourist attraction sites.

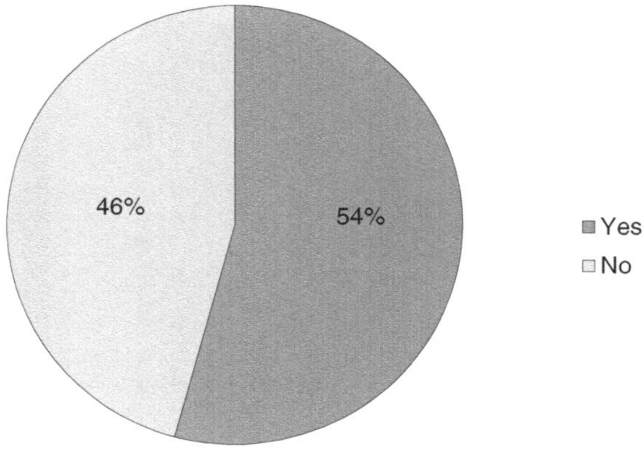

Figure 3.1 Corporate social responsibility increases participation in rehabilitation of tourist sites

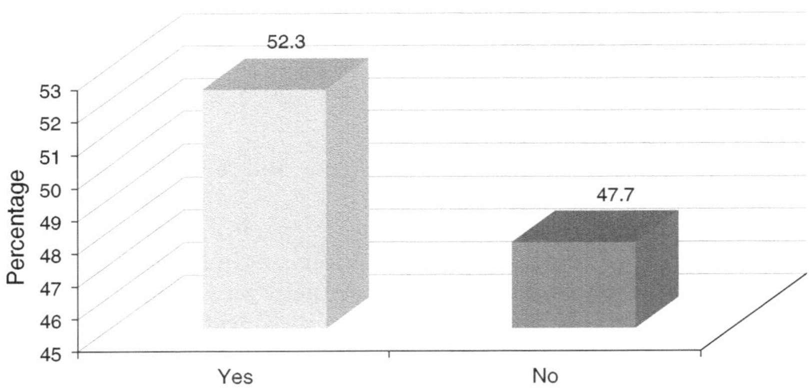

Figure 3.2 Corporate social responsibility helps fund tourism-marketing activities in creating sustainable tourism

Results on whether CSR funds tourism-marketing activities in creating sustainable tourism showed that 52.3% of the respondents were in agreement, while 47.7% were not in agreement with the opinion (Figure 3.2).

Research findings on the importance of CSR in creating sustainable tourism indicated that the majority (39.7%) of the respondents agreed to the opinion that it attracts and retains investors, 30.7% agreed that it increases creativity, 17.8% agreed that it boosts tourism engagement, and 11.8% were of the opinion that it improves public image (Figure 3.3). This implies that through practicing CSR, measures of enhanced tourism sustainability in destinations (in this

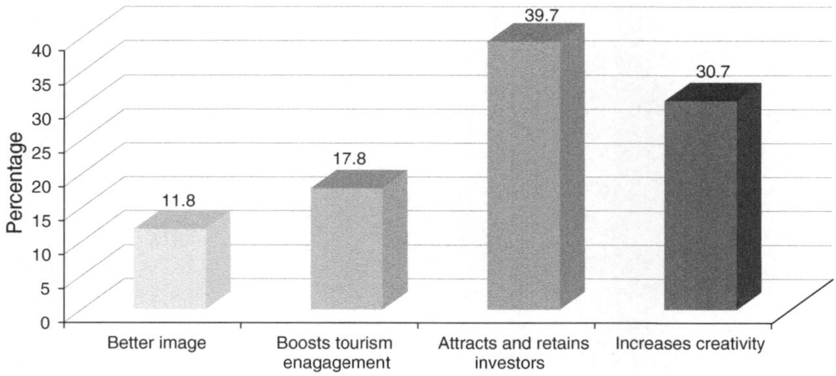

Figure 3.3 Importance of corporate social responsibility in improving sustainable tourism

case—investments creation, public image, engagement, and creativity) are positively influenced.

4.4 Test of Hypotheses

The following was the hypothesis of the study and the conclusion drawn after testing:

H_0: There is no significant effect of tour operators' corporate social responsibility on sustainable tourism development in Nairobi County.

The results of the study indicated that there was a significant effect between CSR and sustainable tourism development (p = 0.000); the study, therefore, rejected the null hypothesis (and accepted its alternative form).

4.5 Discussion of the Interview Schedule

The study employed interview schedules drawn from the constructs/variables of the study objectives. Face-to-face conversations with research respondents—managers of the selected tour-operating firms and KATO representatives—were conducted.

The responses from the interviews demonstrated a direct positive correlation between tourism sustainability and the variables. There was a general observation that tour operators' contribution to societal needs and processes could enhance sustainability of tourism within Nairobi, and by extension to other tourist destinations.

Tour firms have embraced CSR by fulfilling this strategic goal/moral need that they are better placed to deal with than other organizations.

There was also agreement with the observation that CSR could be used by tour operators as a strategic tool if the program is aligned to tour operators' goals, resulting in a competitive edge over others in the same way that price can be used successfully to weave into tour operators' objectives. In this case, a tour-operating company can also become a trendsetter among competitors seeking to use it as a benchmark on corporate social responsibility, hence effect, on sustainable tourism development.

5. Study Limitations

This study was limited by the following factors: the research was self-financed, therefore budget limited; the sample of respondents on which the results were based may have been influenced by personal perceptions, beliefs, and experiences, leading to bias; and questionnaires were written only in English. This last factor was a limitation in that respondents who spoke a different language could not directly communicate in English—translation was required in this case.

6. Conclusion and Recommendations

Based on the background literature and from the findings of the study, it can be concluded that corporate social responsibility is an important aspect in sustainable tourism development and could be used by tourism stakeholders as a tool for gaining competitive edge.

Tourism sustainability should be a big part of a tour operator's (and by extension, other stakeholders in tourism) marketing strategies and relations with its customers, as well as its partnerships and presence in destinations. In enhancing destination sustainability, tour operators could also participate in rehabilitation of tourist attraction sites. This is because tourism is not always a clean business and at times leads to degradation of resources. These responsibilities are now being acknowledged by pioneering tour operators and welcomed by their markets and other stakeholders.

Stakeholders in the tourism sector should collaborate to create practical tools for enhancing sustainability, including action plans and strategies at the local, national, regional, and international levels utilizing multi-stakeholder processes. This would potentially ensure that tourism sustainability is firmly grounded in industry practices and that responsibilities are clearly defined and complementary toward the common societal good.

Finally, the study recommends additional research into other tourism stakeholders, including the local community and government and private actors, among others, to validate these results. There is also a need to test other potential indicators of tourism sustainability other than stakeholder involvement. Nonetheless, the study is an important pacesetter in conceptualizing tour operator roles.

Note

1. Kenya Association of Tour Operators—the official registered body for tour operators in Kenya.

References

Agyeman, J. (2005). *Sustainable communities and the challenge of environmental justice*. New York: NYU Press.

Ahmad, F., Draz, M. U., Su, L., & Rauf, A. (2019). Taking the bad with the good: The nexus between tourism and environmental degradation in the lower middle-income Southeast Asian economies. *Journal of Cleaner Production*, *233*, 1240–1249.

Akama, J. S. (1997). Tourism development in Kenya: Problems and policy alternatives. *Progress in Tourism and Hospitality Research*, *3*(2), 95–105.

Amartya, S. (1999). *Development as freedom*. New York: Random House.

Bauman, W. A., Bohannon, R., & O'Brien, K. J. (Eds.). (2017). *Grounding religion: A field guide to the study of religion and ecology*. New York: Taylor & Francis.

Barry, J. (2007). *Environment and social theory* (2nd ed.). London and New York: Routledge.

Buckley R. (2011). *Understanding and managing tourism impacts: An integrated approach*. *Journal of Ecotourism*, *10*(2), 177–178.

Budeanu, A. (2005). Impacts and responsibilities for sustainable tourism: A tour operator's perspective. *Journal of cleaner production*, *13*(2), 89–97. https://doi.org/10.1016/j.jclepro.2003.12.024

Buhalis, D. (2001). *Tourism distribution channels: Practices, issues and transformations*. London: Thompson Learning Press.

Caton, K. (2016). A humanist paradigm for tourism studies. *Tourism Research Paradigms: Critical and Emergent Knowledges*, 35–56. https://doi:10.1108/s1571-504320150000022009

Cheng, L., & Zhang, J. (2020). Is tourism development a catalyst of economic recovery following natural disaster? An analysis of economic resilience and spatial variability. *Current Issues in Tourism*, 1–22.

Curtin, S., & Busby, G. (1999). Sustainable destination development: The tour operator perspective. *International Journal of Tourism Research*, *1*(2), 135–147.

Cohen, S., Higham, J. E., Peters, P., & Gossling, S. (2014). *Why tourism mobility behaviors must change. Ch. 1 in: Understanding and governing sustainable tourism mobility: Psychological and behavioral approaches*. Cape Town: Juta.

Cistulli, V. (2002). Environment in decentralized development—economic and institutional issues. *Food and Agriculture Organization of the United Nations*. Retrieved from www.fao.org/3/y4256e/y4256e04.htm

Dahiru, M. (2011). New directions in tourism for third world development. In *Annals of Tourism Research*. Princeton, NJ: Van Nostrand.

Daly, H. E. (1996). *Beyond growth: The economics of sustainable development*. Boston, MA: Beacon Press.

DeCarvalho, R. J. (1991). The humanistic paradigm in education. *The Humanistic Psychologist*, *19*(1), 88–104. https://doi.org/10.1080/08873267.1991.9986754

Dent, D. (2012). *International Tourism in the developing countries: New strategy in tourism Pergamon*. Amsterdam Publisher.

Denzin, N. K. (Eds.). (2000). *Handbook of qualitative research*. London: Sage, Lincoln YS.

Drobotova, M., Krasnomovets, V., Radchenko, O., & Romanov, A. (2019). Predicting activity results of the specialized tour operator. In *SHS web of conferences* (Vol. 65, p. 05002). Les Ulis, France: EDP Sciences.

Fan, X., & Thompson, B. (2001). Confidence intervals about score reliability coefficients, please: An EPM guidelines editorial. *Educational and Psychological Measurement*, *61*, 517–531.

Fernando, G. (2007). *Strengthening backward economic linkages: Local food purchasing by three Indonesian hotels. Tourism Geographies*. New York: Columbia University Press.

Fisher, J. (2004). Social responsibility and ethics. *Journal of Business Ethics*, *52*, 391–400.

Fisher, L. D. (1998). Self-designing clinical trials. *Statistics in Medicine*, *17*, 1551–1562.

Font, X., & Cochrane, J. (2005). *Integrating sustainability into business* (Vol. 2). Hertfordshire: UNEP/Earthprint.

Foss, G. (1997). *A vision of tourism for the new southern Africa: Why tourism matters*. Paper prepared for the launch of Action for South Africa's People-First Tourism campaign.

Goldstein, H. (1986). Toward the integration of theory and practice: A humanistic approach. *Social Work*, *31*(5), 352–357.

Gössling, S. (2003). Market integration and ecosystem degradation: Is sustainable tourism development in rural communities a contradiction in terms? *Environment, Development and Sustainability*, *5*(3–4), 383–400.

Gossling, S., Hall, M., Peeters, P., & Scott, D. (2010). The future of tourism: Can tourism growth and climate policy be reconciled? A mitigation perspective. *Tourism Recreation Research*, *35*(2), 119–130.

Gray, R., Kouhy, R., & Lavers, S. (1995). Corporate social and environmental reporting: A review of the literature and a longitudinal study of UK disclosure. *Accounting, Auditing & Accountability Journal*, *8*(2), 47–77. https://doi.org/10.1108/09513579510146996

Griessler, E., & Littig, B. (2005). Social sustainability: A catchword between political pragmatism and social theory. *International Journal for Sustainable Development*, *8*(1/2), 65–79.

Hoffman, A. J. (2000). Integrating environmental and social issues into corporate practice. *Environment: Science and Policy for Sustainable Development*, *42*(5), 22–33. https://doi.10.1080/00139150009604887

Holloway, J. C., Humphreys, C., & Davidson, R. (2010). *The business of tourism* (8th ed.). Harlow: Pearson Education Limited.

Høyer, K. G. (2000). Sustainable tourism or sustainable mobility? The Norwegian case. *Journal of Sustainable Tourism*, *8*(2), 147–160.

Jenkins, W. (2009). Sustainability theory. *Berkshire Encyclopedia of Sustainability: The Spirit of Sustainability*, 380–384.

Kerlinger, F. N (1973). *Foundations of behavioral research* (2nd ed.). New York: Holt, Rinehart and Winston.
Khairat, G., & Maher, A. (2012). Integrating sustainability into tour operator business: An innovative approach in sustainable tourism. *Tourismos: An International Multidisciplinary Journal of Tourism*, *7*(1), 213–233.
Kornienko, A. A. (2017). Ideals and principles of common wellbeing under social and humanistic paradigm. https://doi:10.15405/epsbs.2017.01.49
Kwena, E. (1997). Kenya Fears huge drop in peak season Tourism. *The East African*.
Lavelle, M. J. (2002). *Community-based tourism in the Asia Pacific*. Ontario/ CTC/APEC: School of Media Studies, Humber College.
Lisse, J. (2010). What is the meaning of sustainable tourism? *USA Today*.
Maroto-Martos, J. C., Voth, A., & Pinos-Navarrete, A. (2020). The importance of tourism in rural development in Spain and Germany. In *Neoendogenous Development in European Rural Areas* (pp. 181–205). Cham: Springer.
Mayne, J. (1997). *Participatory planning: A view of tourism in Indonesia*. Oxford: Goodfellow Publishers Limited.
Mensah, J., & Ricart Casadevall, S. (2019). Sustainable development: Meaning, history, principles, pillars, and implications for human action: Literature review. *Cogent Social Sciences*, *5*(1). doi:10.1080/23311886.2019.1653531
McIntyre, G. (1993). *Sustainable tourism development: Guide for local planners*. Madrid: World Tourism Organization (WTO).
Miretpastor, L., Peiró-Signes, Á., Segarra-Oña, M., & Mondéjar-Jiménez, J. (2015). The slow tourism: An indirect way to protect the environment. In H. G. Parsa (Ed.), *Sustainability, social responsibility and innovations in tourism and hospitality* (pp. 317–339). Oakville, ON, Canada: Apple Academic Press.
Mishra, P. K., & Kestwal, A. K. (2019). Tourism-energy-environment-growth nexus: Evidence from India. *Journal of Environmental Management and Tourism*, *10*(5), 1180–1191.
Mitchell, J., & Ashley, C. (2010). Tourism and poverty reduction: Pathways to prosperity. Cambridge and New York: CABI Publishing.
Mugenda, O. M., & Mugenda, A. G. (2003). *Research methods: Quantitative & qualitative approaches*. Nairobi: African Centre for Technology Studies Press.
Niedziółka, I. (2014). Sustainable tourism development. *Regional Formation and Development Studies*, *8*(3), 157–166.
Omollo, J. (2012). Local economic development: Tourism—Good or Bad? In *Tourism workshop proceedings: Small, medium, micro enterprises*. Nairobi: Jomo Kenyatta Foundation.
Patton, M. Q. (2002). Two decades of developments in qualitative inquiry: A personal, experiential perspective. *Qualitative Social Work*, *1*(3), 261–283.
Peeters, P., & Dubois, G. (2010). *Tourism travel under climate change mitigation constraints*. *Journal of Transport Geography*, *18*(3), 447–457. https://doi:10.1016/j.jtrangeo.2009.09.003.
Peeters, P., Gössling, S., Ceron, J. P., Dubois, G., Patterson, T., Richardson, R. B., & Studies, E. (2004). *The eco-efficiency of tourism*. Toronto: Pearson Prentice Hall.
Rangan, K., Chase, L. A., & Karim, S. (2012). *Why every company needs a CSR strategy and how to build it*. Pennsylvania State University Press.
Ribot, T. (2004). *Tourism change, impacts and opportunities*. Toronto: Pearson Prentice Hall.

Ristić, D., Vukoičić, D., & Milinčić, M. (2019). Tourism and sustainable development of rural settlements in protected areas-Example NP Kopaonik (Serbia). *Land Use Policy*, *89*, 104231.

Rogovysky, L., & Dunfee, J. (2002). *Tourism and development in the third world*. London: Routledge.

Scoones, I. (2016). The politics of sustainability and development. *Annual Review of Environment and Resources*, *41*, 293–319. https://doi.org/10.1146/annurev-environ-110615-090039

Sharma, A. K., & Talwar, B. (2005). Insights from practice. Corporate social responsibility: Modern vis-à-vis Vedic approach. *Measuring Business Excellence*, *9*, 35–45. http://dx.doi.org/10.1108/13683040510588828

Smith, R. (2007). *Tourism and development in the developing world*. New York: Routledge.

Solow, R. M. (1993). Policies for economic growth. In *Tinbergen lectures on economic policy* (pp. 127–140). Amsterdam: North-Holland.

Styles, D., Schönberger, H., & Galvez Martos, J. L. (2013). *Best environmental management practice in the tourism sector*. Luxembourg: Publications Office of the European Union.

Thiong'o, S. M. (2019). Insecurity impacts, effects and implications for the Tourism industry—an analysis of selected incidences of terrorism, and safety issues during election years in Kenya. Unpublished Manuscript.

Tøllefsen, I. B. (2011). Grounding religion, a field guide to the study of religion and ecology by Whitney A. Bauman, Richard R. Bohannon II and Kevin J. O'Brien, eds. *Alternative Spirituality and Religion Review*, *2*(2), 379–382. https://doi:10.5840/asrr20112211

UNECOSOC (UN Economic and Social Council). (2008). *Climate change: African perspectives for a post-2012 agreement*. First joint annual meetings of the African Union conference of ministers of economy and finance, and Economic Commission for Africa conference of African ministers of finance, planning and economic development. Addis Ababa, Ethiopia, 26–29 March. UN E/ECA/COE/27/8.

UNDESA (United Nations Department of Economic and Social Affairs). (2007a). *Achieving sustainable development and promoting development cooperation*. New York: United Nations Publications. ISBN: 978-92-1-104587-1

UNDESA (United Nations Department of Economic and Social Affairs). (2007b). *Indicators of sustainable development: Guidelines and methodologies* (3rd ed.). Retrieved from https://sustainabledevelopment.un.org/index.php?page=view&type=400&nr=107&menu=1515

Valls, J.-F., Mota, L., Vieira, S. C. F., & Santos, R. (2019). Opportunities for slow tourism in Madeira. *Sustainability*, *11*(17), 4534. https://doi:10.3390/su11174534

Viewpoint, L. (2008). *Local participation in tourism in the West Indian Islands*. London: Routledge.

Wall-Reinius, S., Ioannides, D., & Zampoukos, K. (2019). Does geography matter in all-inclusive resort tourism? Marketing approaches of Scandinavian tour operators. *Tourism Geographies*, *21*(5), 766–784.

4 Sharing Economy in Peripheral Tourism Destinations

The Case of Finnish Lapland

Petra Paloniemi, Salla Jutila, and Maria Hakkarainen

1. Introduction

Tourism is one of the biggest economic forces in Finnish Lapland. Lapland represents an example of a peripheral destination, where the growth numbers of international tourism have been remarkable for years. Also, the relative share of the sharing economy is significant both on a national and a global level—for example, the number of Airbnb listings per capita is higher in the capital of Lapland, Rovaniemi, compared with Helsinki or even Barcelona. The sharing economy has radically changed the concept of the hospitality and tourism ecosystem. As a relatively new and unpredictable phenomenon, it is crucial to try to understand the impacts of the sharing economy on tourism destinations for the sake of sustainable development and balanced growth. This case study introduces the scope and the implications of the sharing economy for sustainable destination management in Finnish Lapland. The study enhances the balanced growth of tourism in peripheral regions and supports the holistic well-being of local people, tourism businesses, and tourists. The study was implemented within the project Possibilities and Challenges in Peer-to-Peer Accommodation (ERDF).

Tourism has long roots in Lapland. At the beginning of the twentieth century, summertime was the most popular season to travel to Lapland, and the typical tourists were people from southern Finland traveling to Lapland to see the midnight sun and landscape. International tourism started in the 1980s when the first Concorde plane landed in Rovaniemi. From 1984 to 1999, Concorde planes brought more than 9,000 tourists, mostly British, to Lapland and created the base for the current charter flights and tourism (Hakulinen, Komppula & Saraniemi, 2007, pp. 10–12). International tourism was, and still is, highly focused in wintertime and especially at Christmas time. In relation to its population, Lapland gets the largest relative share of tourists in Finland (Business Finland, 2019), and the number of registered international stays has tripled in the 2000s. Today, there are only approximately 181,000 people living in Lapland, but about 2.9 million registered overnights annually. The annual growth rate of overnights is about 9% (House of Lapland, 2019).

Today's tourists seek emotional experiences (Frochot & Batat, 2013), alternative destinations, and accommodation which engage them on a personal level, to indulge in the experiential aspects of consumption (Camilleri & Neuhofer, 2017; Chandler & Lusch, 2015; Holbrook & Hirschman, 1982). In Lapland, the heritage and culture of the people of the north, unspoiled nature and the skill needed to survive in the Arctic environment are attractive dimensions for tourists. The sharing economy offers opportunities to respond to travelers' need to become acquainted with the local way of life, as it enables temporary access to local, everyday life at the destination (Paloniemi, Jutila & Hakkarainen, 2018).

The growth of tourism has also created a need for more and new kinds of services in Lapland, especially in the accommodation sector. Many local people and companies have seen business potential in organizing and offering accommodation and other services (e.g. everyday life experiences with locals and mobility services) for tourists. Online platforms like Airbnb have made it easy for local people to offer peer-to-peer accommodation and experiences to visitors. Along with business potential and many positive opportunities, the changing ecosystem with sharing economy–related services has also caused many legal, social, and economic challenges. The traditional hospitality sector, destination management organizations, and local inhabitants are facing an unprecedented situation as the so-called disruptive force of the sharing economy seems able to create value for tourists and also for locals.

In the twenty-first century, awareness and sensitivity concerning global economic, socio-cultural, and environmental issues have risen, leading to discussion of more sustainable tourism development (Aronsson, 2000; Hall, 2000; Tasci, 2017). The sharing economy can be described as a disruptive force (Guttentag, 2015) which has changed and will probably continue to change the tourism ecosystem in many destinations. It is crucial for tourism destinations to be able to recognize and handle change (Dredge, 2016; Hall, 2013; Ritchie & Crouch, 2003). To enhance common well-being at the destination and avoid conflicts among the different stakeholders—traditional business, local residents, investors, and sharing-economy actors—the changing ecosystem has to be defined and taken into consideration when planning sustainable destination management strategies.

2. Theoretical Framework

2.1 Sharing Economy in Peripheries

The sharing economy can be defined as a marketing practice that links consumers, business and society through virtual networks (Bardhi & Eckhardt, 2012; Belk, 2014; Botsman, 2013; Hakkarainen & Jutila, 2017).

Similar terms include 'collaborative economy', 'peer economy', 'peer-to-peer consumption', and 'access-based economy'. Sharing economy is, however, the most common term to describe a business model based on sharing, even if there has been considerable discussion of the paradox of the sharing economy: how sharing can be considered as economy and how commercial action can be sharing (Arnould & Rose, 2016; Belk, 2014; Hakkarainen & Jutila, 2017). Commercialization of the sharing economy has brought different kinds of challenges in many destinations, especially related to peer-to-peer accommodation. Property prices and rents have risen in popular tourist cities, and local inhabitants face difficulties in finding apartments near the city center (Lambea Llop, 2017). The share of Airbnb capacity is also considered an indicator of overtourism (Peeters et al., 2018).

Despite the high density of peer-to-peer accommodation, the sharing economy is often considered a positive phenomenon, especially in the peripheral regions. The situation is very different compared with the situation in big European cities where there are heated debates concerning regulations, for instance, related to taxation, safety, and disturbance in the neighborhood around the sharing economy. Also, many traditional tourism businesses and local inhabitants feel threatened by the rise of sharing-economy services. Finland operates in a national regulatory environment; regions and cities follow the national framework. As previously stated, Finland, and especially Finnish Lapland, are examples of peripheral regions, where the sharing economy can be expected to sustain livelihood in the rural surroundings (Battino & Lampreu, 2019). Sometimes, from the individual perspective, the notion of dignity represents a missing link to the quest for social welfare (Pirson, 2017).

In tourism, the sharing economy, on the one hand, emphasizes individual experiences, along with a sense of communality and belonging (Bock, 2015; Hakkarainen & Jutila, 2017). It offers various opportunities to bring the features of local lifestyles to tourists as it enables temporary access to local services, products and everyday life (Belk, 2014, p. 1595). Many studies have shown that sharing-economy users are driven by a willingness to have local-like experiences at the destination (Gansky, 2010; Guttentag, 2015; Sigala, 2017; Tussyadiah & Pesonen, 2018; Volgger, Pforr, Stawinoga, Taplin & Matthews, 2018). According to Camilleri and Neuhofer (2017), a number of guests pointed out that the need to seek a quiet place for relaxation, and to avoid 'tourist locations' for the enjoyment of a more authentic experience of the setting, was a reason for choosing accommodation via the sharing-economy platform (Airbnb in this case). The quest for real-life experiences has also been noted in various national and regional strategic guidelines, for instance, in Lapland, Copenhagen, Helsinki, Amsterdam, and Iceland. On the other hand, the sharing economy is considered in tourism as a commercial activity and as a business opportunity. Through the sharing economy, individuals living, for example, in peripheral regions can take part in the market economy and even employ themselves (Huefner, 2015). The sharing economy enables an active role for local people, who often are positioned as objects of tourism development, not as active subjects participating in tourism. This supports the perspective of humanistic management emphasizing the need to understand human beings throughout the economic system as active subjects instead of being treated merely as passive objects (Dierksmeire, 2016, p. 25).

There is only limited research regarding the sharing economy in peripheral destinations. The term 'peripheral' is often used to refer to a region that is remote and sparsely populated, and that is characterized by a lack of public and private services, and limited in accessibility (Brown & Hall, 2000). Over the years, tourism has become a development catalyst for promoting social and economic welfare in peripheral areas and their communities (Brown & Hall, 2000; Saarinen, 2010; Sharpley, 2015). As Battino and Lampreu (2019, p. 3) argue, in rural areas, sharing-economy practices offer solutions and alternatives to the classic models of hospitality that are often absent in peripheral areas. Finnish Lapland as a whole can be seen as a peripheral region. Within the region, there are different kinds of peripheries, where the impacts of the sharing economy are also different. Rovaniemi, the largest town in Lapland with around 62,000 inhabitants, represents an urban city in the periphery, facing similar challenges related to the sharing economy to those faced by the big cities of central Europe. Remote villages and skiing resorts in Lapland can be considered as peripheries within the periphery. For these districts, the sharing economy enables easy and cheap business and marketing possibilities, as well as new kinds of income opportunities. As the sharing economy does not require large investments, it can cause rapid and impressive changes to the structures of tourism ecosystems, especially in peripheral destinations. It is possible to assume that many tourists are willing to spend their holidays in peaceful and authentic peripheral regions (Camilleri & Neuhofer, 2017), like small villages in Lapland, in which sometimes they can only find accommodation and activities through peer-to-peer services. Thus, the sharing economy might widen the tourism spectrum and spread the impacts of tourism to more peripheral areas.

2.2 *Sustainable Destination Management*

Sustainability is an important principle in the management and competitiveness of tourism destinations (D'Angella & De Carlo, 2016; Middleton & Hawkins, 1998; Mihalic, 2000; Ritchie & Crouch, 2003; Volgger, Pforr, Stawinoga, Taplin & Matthews, 2018) and is a very topical issue at the moment. In October 2018, the Intergovernmental Panel on Climate Change (IPCC) report was launched, attracting worldwide attention. The IPCC reported on the impacts of global warming of 1.5°C above pre-industrial levels and related global greenhouse gas emissions pathways, in the context of strengthening the global response to the threat of climate change, sustainable development, and efforts to lessen poverty (IPCC, 2018). The report has provoked discussion widely in many tourism-related forums. The report, and the ensuing discussions, will probably cause changes in actions both at the individual and national levels, which might lead to impacts on tourism and tourism behavior. It is reasonable to assume that in the future, sustainability issues will become even more and more important to the success of destinations.

Sustainable development can be defined as 'development that meets the needs of the present without compromising the possibilities of the future generations' (WCED, 1987, p. 43). In tourism research, management of the impacts and governance of tourism development toward sustainability has already been an important issue for years (Saarinen, 2014; Tervo-Kankare, Kaján & Saarinen, 2018). Tourism is affected by, and also a reason for, changes taking place in physical, environmental, sociocultural, or economic environments. This creates the need for sustainability management (Bramwell & Lane, 2011; Hall, 2013).

'Destination' here may refer to a relatively coherent spatial unit that includes tourism and tourism-related businesses and other actors collaborating in co-production, in local value chains, and often also in marketing (Saarinen, 2004). Tourism destinations compete against one another for potential visitors. According to Ritchie and Crouch, 'a tourism destination is competitive when it has got an ability to increase tourism expenditure, to attract the visitors while providing them with satisfying, memorable experiences in profitable way while simultaneously enhancing the wellbeing of the destination residents and preserving the natural capital of the destination for future generations' (2003). This definition emphasizes the well-being and recovery of the everyday lives of local people at the destination, which also creates better opportunities for positive visitor experiences. The definition also relates to the conception of humanistic management, that according to Dierksmeire (2016, p. 28) encompasses moral, social, and ecological sustainability criteria as it considers individual autonomy as an obligation for self-commitment: individual liberty and common responsibility belong together and strengthen each other. Thus, humanistic management aims to use business as a way to improve every individual's conditions in the interest of their dignity (Dierksmeire, 2016, p. 28).

At many destinations, the tourism destination strategies are planned by DMOs. DMO may refer either to 'destination marketing organization' or 'destination management organization', depending on the organization's activities. Both of them are policy tools to stimulate tourism growth (Dredge, 2016). Some researchers doubt that DMOs even fit in an increasingly liquid and mobile world (Bauman, 2000; Dredge, 2016). Dianne Dredge calls for DMOs to take social responsibility, stewardship, and sustainability into account, as she sees these as important values, especially in future destinations. Dredge assumes that in the future there will be more demanding challenges as tourism moves from a formal, industrial structure to a post-structural organization in which increased mobilities, and the increasing use of social media and collaborative economy, is common (Dredge, 2016).

2.3 Sustainability in Sharing Economy–Based Tourism Management

In their literature review, Nuottila, Jutila and Hakkarainen (2017) noticed that, despite the common interest in a sharing economy, there are only a limited number of studies regarding the increasing diversity of the sharing economy and its implications for sustainability. The sustainable idea beneath sharing economy is that it is better to 'share rather than own', as this leads to a sustainable decrease in personal consumption, ownership, and waste (Sheth, Sethia & Srinivas, 2011). The sharing economy holds the promise for a more sustainable world by giving access to underutilized resources at a small cost to those who cannot or do not want to buy new products or facilities, and for those who already own such resources, there is also a chance of making extra income. Airbnb's mission is 'to create a world where people can belong through healthy travel that is local, authentic, diverse, inclusive and sustainable' (Airbnb, 2018a). The sharing economy also enables services to be offered in remote destinations where conventional businesses might not be profitable.

Nevertheless, there are debates concerning regulations (e.g. taxation, gray market, safety, disturbance in the neighborhood) around the sharing economy in tourism. At the same time of having plenty of sharing economy start-ups, new entrepreneurs, and investors, there also exists fear among many other stakeholders who feel threatened by the sharing economy, for example, traditional tourism businesses and local inhabitants. According to Williams and Horodnic (2017), businesses in the hotel and restaurant sector see the sharing economy, considered an informal sector in business, as an obstacle or even a threat to their own business. Often, the sharing economy is also seen as instrumental in facing problems such as overconsumption and income inequality. Sharing economy–based businesses have evolved from simple peer-to-peer lending initiatives to complex platforms and networks of people and companies interacting for the collective use of new or extant resources. According to Muñoz and Cohen (2017), most of the current sharing-economy brands have been supported by mainstream investors, moving business models away from the still-predominant view of sharing businesses as driven by social-oriented goals. It is expected that regulations of some kind will see daylight in different countries and maybe also at the EU level in the years ahead. Although regulations may reduce the sharing economy, it is unlikely that it would be prohibited, its growth reversed, or that its business logic would vanish totally.

Social responsibility is one key issue that must be taken into consideration when developing tourism experiences based on the sharing economy. In some European cities, sharing economy–based tourism services have grown explosively without sustainable planning, causing problems in local communities. It is thus essential to note that sharing economy–based services have to be developed on locals' terms, from the local perspective (Jutila, Paloniemi & Hakkarainen, 2017).

Dredge and Gyímothy (2015) state that traditional DMOs are being challenged by the rise of the sharing economy. Peer-to-peer exchange of tourist goods and services (e.g. Airbnb, Doerz, Blablacar, etc.) are not only challenging the traditional incomes of DMOs but also making tourism information services redundant. As a result, the traditional business models of DMOs are under pressure (Dredge & Gyímothy, 2015). Also, social media is challenging traditional destination management structures. In the analysis of social media initiatives in Denmark, Norway, Finland, Sweden, and the Scandinavian Tourist Board, Munar found that the fluid, decentralized, and user-generated characteristics of social media make it challenging for DMOs to control and manage marketing and branding efforts (Dredge, 2016; Munar, 2011, 2012). Trunfio and Della Lucia (2019a, 2019b) call for the potential of stakeholder engagement for value co-creation in destination strategy. According to the authors, digitalization may itself drive this kind of development, for instance, by facilitating interactive communication (Trunfio & Della Lucia, 2019a, p. 202).

Sharing economy services can, however, facilitate access to insider explanations about the destination, which is particularly precious in cases of low destination familiarity (Volgger, Pforr, Stawinoga, Taplin & Matthews, 2018). In addition, sharing economy services are often cheaper, and they may boost first visits to destinations by reducing cost barriers that previously hindered the willingness to visit the place. The existence of a sharing economy may thus create competitive advantage for certain destinations. According to Airbnb statistics, 42% of guest spending happens in the neighborhoods where guests stay (Airbnb, 2018b), so sharing economy services may create benefits to neighborhood areas in other sectors as well and spread positive economic impacts. Toni, Renzi and Mattia (2018) found out that collaborative consumption stimulates sustainable behavior in travelers. Their study investigated the relationship between collaborative consumption and the adoption of sustainable practices from the users' angle. They found that it is necessary to emphasize the importance of defining a new business paradigm that considers collaborative consumption as a way of supporting environmental policy and encouraging sustainable behavior.

The growing controversy around the adverse impacts of sharing economy platforms has led to calls for more democratic models of platform governance. This could help to create a more sustainable sharing economy by ensuring that platforms promote social and environmental values alongside the instrumental values of the capitalist economy. There are already existing platforms that, for example, enable people to gift unwanted consumer goods locally (Martin, Upham & Klapper, 2017). Noirbnb and Fairbnb are examples of platforms which promote community-centered alternatives, prioritize people over profit and create travel experiences. For example, Fairbnb 'aims at putting the "share" back into the sharing economy' and has got collective ownership and democratic governance, and they promote 'social sustainability, transparency and accountability' (Fairbnb, 2019; Noirbnb, 2019.) It can be easily assumed that multiple similar platforms will come to light soon, as many stakeholders are interested in more sustainable and democratic values.

3. Case Study Method

In this study, we have used a case study approach. That approach was chosen to gain a deep understanding of the phenomenon in a local context with global frames (Yin, 2014). This case can provide important new insights into the relatively new but recently actively researched tourism trend of sharing economies. The strong growth of the phenomenon is undoubtedly attached to urbanization and has also required a sufficient number of both users and suppliers to grow to the current scale. However, as already mentioned, the phenomenon not only exists in big cities and large population centers but also affects more peripheral and sparsely populated areas. In our research, we have brought the periphery into focus.

Based on our previous studies and development projects, we have noticed that the main field of the sharing economy in Lapland and in Rovaniemi is peer-to-peer accommodation (Hakkarainen & Jutila, 2017). Previous studies of peer-to-peer accommodation have focused primarily on explanations and descriptions of the phenomenon on a general level (Bardhi & Eckhardt, 2012; Belk, 2014; Botsman, 2013); on the platforms, especially Airbnb's role in accommodation business (Guttentag, 2015); on the drivers to choose peer-to-peer accommodation (Tussyadiah & Pesonen, 2018); on trust and hospitality (German-Molz, 2013); and on the social and legal challenges the phenomenon has caused, especially in popular tourism destinations and big cities (Dredge, 2016). Thus, our attempt is to broaden the discussion to include a topic which has gotten little attention.

In the data collection, we have used data triangulation (Laine, Bamberg & Jokinen, 2007): our data consist of both statistical information and qualitative data. We have collected statistical data from Finnish national statistics (Statistics Finland), regional statistics from Lapland

(Council of Lapland; House of Lapland) and statistics from Airbnb (AirDNA.co), including tourism and regional development indicators: volumes, utilization rate, prices, seasons, map-based data of the region, and the number of citizens. Qualitative data are gathered from several sources (interviews, newspapers, websites, focus groups, and project workshop observations), websites of sharing economy platforms (e.g. Airbnb, Doerz, and BloxCar), and other accommodation platforms used in commercial peer-to-peer accommodation, for example, booking.com.

Interviews and focus group discussions with different stakeholders were conducted in Shareable Tourism (ERDF) and Possibilities and Challenges in Peer-to-Peer Accommodation (ERDF) projects. One special source of data is a free distribution pamphlet, which was delivered to almost every household in Rovaniemi in May 2019 (eight color pages in the form of a local newspaper). In our analysis, we have used content analysis as a method and combined qualitative and descriptive quantitative methods (Elo & Kyngäs, 2008). Through statistical information, we have illustrated and mapped both absolute and relative volumes and scale of the sharing economy in tourism in Lapland and in Rovaniemi. Thus, through our case study, we could provide a novel means for understanding different aspects of the sharing economy, which can then be applied to practice.

4. The Sharing Economy in Lapland and Rovaniemi

4.1 Sharing Economy–Based Services in Lapland

As part of our study, we have mapped the most important sharing economy–based services in Lapland. The growth of sharing economy services has been in line with the growth and seasonality of tourism in general in Rovaniemi. A good example is the development of Airbnb listings in Rovaniemi (Figure 4.1).

In March 2016, there were only 136 Airbnb listings in Rovaniemi (Hakkarainen & Honkanen, 2017, p. 31). In June 2017, nearly 400 listings were available. A year later, the number was about 550, and in December 2019, already more than 1,000 (Airdna, 2019). The growth is remarkable, and the number is significant, especially in relation to the number of inhabitants in Rovaniemi, which is around 62,000. In January 2019, there were 14.4 Airbnb listings per 1,000 inhabitants in Rovaniemi, whereas in Helsinki, the capital of Finland, the number was only 4.2. Even Barcelona, Spain, one of the cities struggling with Airbnb and trying to find ways to regulate it, had fewer properties per capita compared with Rovaniemi (Table 4.1).

Almost half (43% in January 2019) of the hosts in Rovaniemi were multi-listing hosts, meaning that they had more than one location listed on Airbnb. Most locations in the center are apartments with one

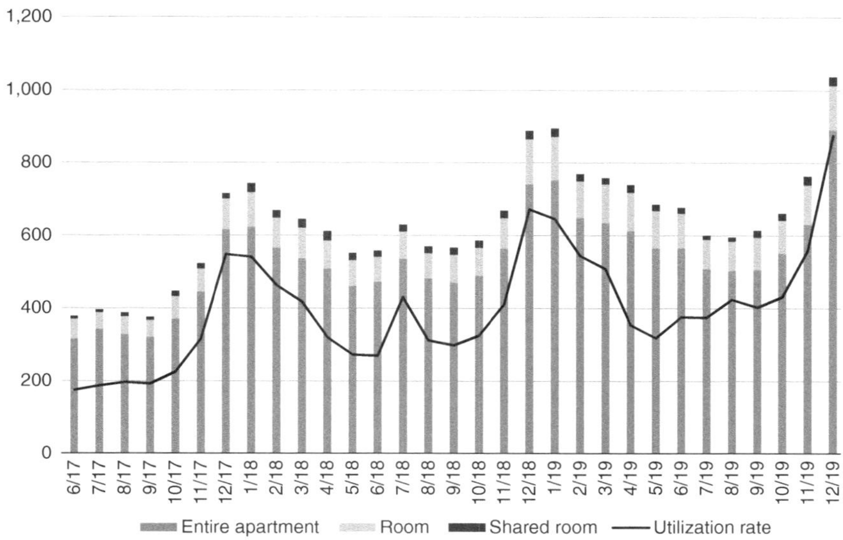

Figure 4.1 Supply and utilization rate of Airbnb listings in Rovaniemi

Table 4.1 Number of Airbnb listings per capita in different cities (January 2019)

Cities	*Listings*	*Listings/1,000 inhabitants*
Rovaniemi	896	14.4
Helsinki	2,739	4.2
Barcelona	20,052	12.5

Source: Authors' own elaboration

bedroom, and in most cases, the owner does not live in the apartment. In January 2019, the monthly revenue average per listing in Rovaniemi was more than in many big European cities (Table 4.2). Annual revenues by top properties, more than 70,000 € (Airdna, 2019), show that peer-to-peer accommodation is a business for many, not just an occasional action. According to this data, we can assume that peer-to-peer accommodation in Rovaniemi follows the same trend as with large tourist cities in Europe: it is changing to be a more business-oriented activity, where the experience is not that personal and authentic anymore but is nevertheless happening among local inhabitants.

In Lapland, the sharing economy phenomenon is, at the moment, strongly centered in Rovaniemi, but the number of different sharing

Table 4.2 Monthly revenue average from Airbnb accommodation in different cities (January 2019)

Cities	*Revenue average*
Rovaniemi	1,541 €
Amsterdam	1,519 €
Paris	1,424 €
Barcelona	1,332 €
Berlin	900 €

Source: Authors' own elaboration

economy services is growing also elsewhere in Lapland. For instance, in Levi, a popular skiing destination in Finnish Lapland, 189 flats were registered on Airbnb and more than 300 homes were registered on the Home Away platform in spring 2019. However, the situation is different in many other remote areas, where the number of properties on Airbnb and other platforms is much smaller, and while the most popular rental size in Rovaniemi is a one-bedroom apartment, listings in remote areas are mostly cottages or private rooms in a house. Because private cottages (second homes) have already been rented through Finnish cottage intermediator companies for decades, the change caused by peer-to-peer platforms is not as remarkable in remote areas as it is in Rovaniemi. However, the selling channels are more versatile nowadays.

Even though accommodation is clearly the biggest sharing economy sector in Lapland, there is also a growing interest in experiences offered by locals. Platforms such as Airbnb Experiences and Doerz have made it easy for locals to offer experiences for tourists. In 2017, there were no experiences existing in sharing economy platforms, but in January 2019, Rovaniemi hosts offered 16 different kinds of activities. The most typical experiences are searching for the northern lights, snowshoeing, road trips to national parks, ice fishing, or Lappish gastronomy experiences. Doerz's platform offers very similar experiences to travelers in Rovaniemi. In spring 2019, Doerz offered 26 activities in Rovaniemi provided by nine hosts; the selection of experiences is more varied compared with Airbnb Experiences. The most common activities are husky safaris, canoeing, aurora hunting, fell hikes, sauna experiences, gastronomy tastings, and reindeer farm visits. The services are provided by both individual local people and small tourism companies. It is obvious that those small tourism companies use the platforms as marketing and selling channels. Doerz is also one of the partner companies of Visit Rovaniemi, the DMO of Rovaniemi.

Sharing economy transportation has also emerged in Lapland but is still quite marginal. BloxCar is a peer-to-peer platform in which car owners can rent their vehicles to other people. In spring 2019, only four cars were registered on the platform in Rovaniemi and single cars elsewhere in Lapland. However, the need for different kinds of sharing practices in transportation is evident, as distances are long and public transportation is poor in Lapland compared with more densely populated areas. Peripheral context has to be taken into account to create successful peer-to-peer transportation. Some housing cooperatives of new blockhouses in Rovaniemi are currently developing such good practices, as houses are planned with car-sharing possibilities.

It can be assumed that the supply of sharing economy–based services will be more diverse in the future. There are plenty of peer-to-peer platforms that are popular globally but are not currently available in Rovaniemi: for example, BlaBlaCar, Glovo, Delivery, Eat With, Uber, Cabify, and Good Timing. The case of Rovaniemi also underlines that the scale of different stakeholders in a sharing economy is versatile. Service providers, users, and platforms are directly related to the phenomenon. In addition to that, there are several stakeholders who are indirectly affected by and influential to different sectors of the sharing economy. Housing cooperatives, steward offices, insurance companies, traditional accommodation businesses, and building supervision are all examples of relevant stakeholders in peer-to-peer accommodation. Based on our case study, Rovaniemi seems to be a living lab in tourism, with sharing economy services and platforms in a national but also in a global context. Furthermore, many new innovations and business ideas have been developed around peer-to-peer accommodation. Different kinds of host, key, cleaning, and concierge services are nowadays sold for apartment owners, who do not have the time or ability to manage everything by themselves. If the number of listings continues to grow in the future, the demand for this kind of ancillary service may also grow.

5. Possibilities and Challenges of a Sharing Economy in a Peripheral Context

As described earlier, the sharing economy in Lapland is a remarkable and versatile phenomenon combining many different stakeholders and fields. Thus, the impacts of the phenomenon are also diverse. Lapland as a whole can be seen as a peripheral region. Within the region, there are different kinds of peripheries, where the impacts of the sharing economy are also different. Rovaniemi represents an urban city on the periphery, facing similar sharing economy–related challenges to those faced by big cities in central Europe. Remote villages and ski resorts in Lapland can, for their part, be considered as peripheries on the periphery. In these districts, the sharing economy largely acts as an enabler of tourism marketing and businesses (see also Huefner, 2015) as sharing economy platforms literally bring villages to the world map. In Rovaniemi, area

Airbnb listings are mostly located in the city center; however, there are several spots all over the periphery of Rovaniemi. Before the appearance of this kind of accommodation possibilities, these places had been outsiders in tourism development and benefits. As large and dominant tourism companies have mainly taken responsibility for tourism development and tourism business in Lapland in places relevant for them, sharing economy platforms enable participation for small actors and individual people anywhere in the periphery.

Even if most Airbnb locations are in the center, accommodation is also available in the suburban areas. Thus, sharing economy–based services might benefit the destination by balancing the pressure and crowding of tourism in the city centers. Because tourists want to 'live like locals' in the neighborhoods (see e.g. Chandler & Lusch, 2015), visit local shops, use the playgrounds in the suburbs, meet ordinary people, and use public transport, this also balances the economic impacts of tourism among different districts. Another goal in Lapland's tourism strategy has been to lengthen the stays of visitors as it is beneficial for the destination in many ways (e.g. for economic, environmental, social, and cultural reasons). In many destinations, tourists who use peer-to-peer accommodation stay longer compared with hotel guests (Airbnb, 2018b). Longer stays also offer better possibilities for a more authentic tourism experience and more contact with the local people, culture, and environment.

Moreover, greater distribution of tourism may prevent some of the negative aspects of growing tourism. Especially during the high season in December, the growing number of tourists in the city center of Rovaniemi causes problems such as crowding, taxi queues, queues in grocery stores, and excessive carrying capacity of public services (e.g. hospitals and police). Also, when tourists are the dominant group in the street view, it might adversely affect the destination's image. There are already some tour operators who do not want to bring their customers to Rovaniemi, but to more 'authentic' and remote destinations in Lapland.

Along with business potential, innovations, and many opportunities, sharing economies inevitably bring different kinds of challenges. The essential challenges in Lapland relate to social disturbance, taxation, gray market, and safety. The central question in all of these is: who is the responsible stakeholder? Even though the challenges are somewhat universal, peripheral regions have particular challenges. The nature of the challenges also differs according to the stakeholder in question and the type of sharing economy service.

Our case study uncovered that many housing cooperatives in Lapland, especially in Rovaniemi, are facing challenges different than ever before, because of peer-to-peer accommodation. Some houses host up to a quarter of the rentals on Airbnb and in other peer-to-peer accommodation use, even though the houses are planned for permanent living. Neighbors are not willing to act as tourist information officers in their own homes.

Continuously changing guests disturbs permanent inhabitants in many different ways: elevators become worn and seedy quickly because of continuous dragging of heavy luggage, house rules are not known and thus not followed by guests who do not speak Finnish and, in some cases, even English, and so on. Tourists who are not familiar with winter conditions also cause many risks when they stay in a house without reception or information. In some cases, tourists have left windows or doors open for hours, because of ventilation or to have a better chance to see the northern lights. In bitterly cold winter conditions, this means a frigid apartment after a while and may also cause frozen heaters. Many rental apartments also have saunas, which can cause their own risks.

Many hospitality entrepreneurs in Lapland see peer-to-peer accommodation providers as a significant threat to the growth of the legitimate hospitality industry. As Williams and Horodnic (2017) point out, the practices of competitors in the informal sector are seen as an obstacle among hotels and restaurants, making, for example, peer-to-peer accommodation in the hospitality industry an issue that needs to be addressed. The contradiction is visible in Lapland in various ways. This has especially become an issue in the Rovaniemi area, as Airbnb has more or less turned the meaning of ownership upside down from the initial principles of the sharing economy: the more apartments you own, the more you can share. The free distribution pamphlet against peer-to-peer accommodation was paid for by local hospitality entrepreneurs, who want to remain anonymous. Traditional accommodation businesses state that they need to apply for several licenses and permissions from different authorities before being able to start an accommodation business in Finland. They view the situation as unfair, because the regulatory environment is not clear regarding peer-to-peer accommodation. There is also concern about taxation; a notable share of income through peer-to-peer accommodation is not declared. The pamphlet, that was implemented in a very provocative style, is a specific example of the pressure the sharing economy has placed on the local tourism ecosystem.

Interestingly, both the traditional accommodation sector and housing co-operatives often blame public authorities and decision-makers for irresponsibility and lack of capability to intervene when problems arise in peer-to-peer accommodation. According to our case study, building supervision in particular faces many requirements to monitor, intervene, and even stop disturbing or somehow unfair peer-to-peer accommodation. Responsibility on the part of hosts has drawn much less attention.

6. Discussion

Sustainability issues are essential in order to develop tourism that promotes well-being. As the national tourism board, Visit Finland, states, sustainable tourism is committed to having a positive impact on nature, society, and the economy, leaving a small ecological footprint and honoring local cultures; to keeping nature clean by choosing environmentally friendly options in modes of travel; and recycle, reuse, and reduce overall consumption and waste. Visit Finland has also set an ambitious goal for Finland to become the most sustainable destination in the world (Visit Finland, 2019).

Dianne Dredge (2016) argues that calls for destination management organizations (DMOs) to embrace social responsibility and sustainability tend to fail often, as, despite societal and political changes, DMOs remain locked within an industrial policy paradigm. Dredge presents a holistic networked approach as a solution to adopt. According to this idea, destination management could be a network of actions, a co-created product of social life and a citizen service, rather than being directed first and foremost at industry interests. This opens up the possibility of re-democratizing tourism management, moving it beyond the confines of an industry-coordination problem (Dredge, 2016). According to Melé (2003), a humanistic management approach fosters both the human growth of people and, as a consequence, their sense of service and cooperation which indubitably are crucial for long-term outcomes. Pirson (2017) emphasized the role of human dignity and the notion of social welfare as either wealth or well-being.

Our case study has shown, that sharing economy–based tourism services create the potential to advance developments for their part. It seems that the sharing economy offers possibilities for individual people to participate in tourism, employ themselves, meet new people from the other side of the world and gain trust in business. The sharing economy also has the potential to enhance dignity through tourism; there is something so valuable and interesting in one's everyday life, that tourists are willing to pay to for the possibility to see and participate in it. Thus, it could be argued, that sharing economy has potential to bring humanism to business (see Dierksmeire, 2016, p. 26).

To enhance common well-being at the destination and avoid conflicts among the different stakeholders, the changing ecosystem has to be defined and taken into consideration when planning sustainable destination management strategies. We have scanned the sharing economy–based services in Lapland, but as a result of statistical restrictions and rapid changes in sharing economy services, we have managed to give only a descriptive estimation of the scale of the phenomenon. To produce more specific and reliable data regarding the volume and the impact of the sharing economy, more reliable statistics should be collected by national or regional administrations. Also, to maximize the benefits and to avoid the challenges caused by the sharing economy, holistic co-creation and stakeholder engagement are needed. Based on our study, it is possible to claim that to avoid extreme reactions (such as the pamphlet against Airbnb), neutral actors who actively advance interaction among different stakeholders, and understand and promote diverse viewpoints, are needed.

According to our research, central challenges to be pondered in further studies of the sharing economy should consider the definition of professional and occasional actors, as well as the difference between accommodation and renting. It is also essential to ask how to make the accommodation sector equal for different stakeholders, how to avoid disturbing local inhabitants, and how to make different responsibilities clear to all stakeholders. The case study from Finnish Lapland presents the current situation of tourism and the sharing economy in a peripheral region and contributes to sustainable destination management supporting the holistic well-being and dignity of local people, tourism businesses and tourists. The project Possibilities and Challenges in Peer-to-Peer Accommodation (ERDF) is one example of sustainable and responsible discussion forums bringing different viewpoints and stakeholders together.

References

Airbnb. (2018a). *About us*. Retrieved from http://airbnb.com/about/about-us

Airbnb. (2018b). *Airbnb economic impact*. Retrieved from http://blog.atairbnb.com/economic-impact-airbnb

Airdna. (2019). Retrieved from www.airdna.co/

Arnould, E. J., & Rose, A. S. (2016). Mutuality critique and substitute for Belk's 'sharing'. *Marketing Theory*, *16*(1), 75–99.

Aronsson, L. (2000). *The development of sustainable tourism*. London and New York: Continuum.

Bardhi, F., & Eckhardt, G. M. (2012). Access-based consumption: The case of car sharing. *Journal of Consumer Research*, *39*(4), 881–898.

Battino, S., & Lampreu, S. (2019). The role of the sharing economy for a sustainable and innovative development of rural areas: A case study in Sardinia (Italy). *Sustainability*, *11*(11), 3004. doi:10.3390/su11113004

Bauman, Z. (2000). *Liquid modernity*. Cambridge: Polity Press.

Belk, R. (2014). You are what you can access: Sharing and collaborative consumption online. *Journal of Business Research*, *67*(8), 1595–1600.

Bock, K. (2015). The changing nature of city tourism and its possible implications for the future of cities. *European Journal of Futures Research*, *3*(20), 1–8.

Botsman, R. (2013). The sharing economy lacks a shared definition. *Fast Company*. Retrieved from www.fastcoexist.com/3022028/the-sharing-economy-lacks-a-shared-definition

Bramwell, B., & Lane, B. (2011). Critical research on the governance of tourism and sustainability. *Journal of Sustainable Tourism*, *19*(4–5), 411–421.

Brown, F., & Hall, D. (2000). Introduction: The paradox of peripherality. In F. Brown, D. D. Hall, & D. R. Hall (Eds.), *Tourism in peripheral areas: Case studies* (pp. 1–6). Clevedon: Channel View Publications.

Business Finland. (2019). *Tourism in Finland stays on record level*. Retrieved from www.businessfinland.fi/en/whats-new/news/2019/tourism-in-finland-stays-on-record-level/

Camilleri, J., & Neuhofer, B. (2017). Value co-creation and co-destruction in the Airbnb sharing economy. *International Journal of Contemporary Hospitality Management*, *29*(9), 2322–2340.

Chandler, J. D., & Lusch, R. F. (2015). Service systems: A broadened framework and research agenda on value propositions, engagement, and service experience. *Journal of Service Research*, *18*(1), 6–22.

D'Angella, F., & De Carlo, M. (2016). Orientation to sustainability and strategic positioning of destinations: An analysis of international tourism websites. *Current Issues in Tourism*, *19*(7), 624–633.

Dierksmeire, C. (2016). What is 'humanistic' about human management? *Humanist Management Journal*, *1*, 9–32.

Dredge, D. (2016). Are DMOs on a path to redundancy? *Tourism Recreation Research*, *41*(3), 348–353.

Dredge, D., & Gyimóthy, S. (2015). The collaborative economy and tourism: Critical perspectives, questionable claims and silenced voices. *Tourism Recreation Research*, *40*(3), 286–302.

Elo, S., & Kyngäs, H. (2008). The qualitative content analysis process. *Journal of Advanced Nursing*, *62*(1), 107–115.

Fairbnb. (2019). Retrieved from https://fairbnb.coop/about-us/

Frochot, I., & Batat, W. (2013). *Marketing and designing the tourist experience*. Oxford: Goodfellow.

Gansky, L. (2010). *The mesh: Why the future of business is sharing*. New York: Penguin Books.

German-Molz, J. (2013). Social networking technologies and the moral economy of alternative tourism: The case of coachsurfing.com. *Annals of Tourism Research*, *43*, 210–230.

Guttentag, D. A. (2015). Airbnb: Disruptive innovation and the rise of an informal tourism accommodation sector. *Current Issues in Tourism*, *18*(12), 1192–1217.

Hakkarainen, M., & Honkanen, A. (2017). Vertaismajoituksen tilastointia: AIRBNB Rovaniemellä. In H. Ilola, P. Satokangas, & M. Tapaninen (Eds.), *Tilastoja tutkimassa: lukuja Lapin matkailusta* (pp. 31–32). MTI: n julkaisuja. Rovaniemi, Finland: Erweko.

Hakkarainen, M., & Jutila, S. (2017). Jakamistalous matkailussa. In J. Edelheim & H. Ilola (Eds.), *Matkailututkimuksen avainkäsitteet* (pp. 183–187). Rovaniemi, Finland: Lapland University Press.

Hakulinen, S., Komppula, R., & Saraniemi, S. (2007). *Lapin joulumatkailutuotteen elinkaari Concorde-lennoista laajamittaiseen joulumatkailuun*. MEK. Helsinki, Finland: Pramedia Oy.

Hall, C. M. (2000). *Tourism planning: Policies, processes and relationships*. Harlow: Prentice Hall.

Hall, C. M. (2013). Framing behavioural approaches to understand and governing sustainable tourism consumption. *Journal of Sustainable Tourism*, *21*(7), 1091–1109.

Holbrook, M. B., & Hirschman, E. C. (1982). The experiential aspects of consumption: Consumer fantasies, feelings, and fun. *Journal of Consumer Research*, *9*(2), 132–142.

House of Lapland. (2019). *10 faktaa Lapin matkailusta*. Retrieved from www.lapland.fi/fi/business/faktat-ja-tilastot/infograafi-10-faktaa-lapin-matkailusta-2018/

Huefner, R. J. (2015). The sharing economy: Implications for revenue management. *Journal of Revenue and Pricing Management, 11*(4), 296–298.

IPCC. (2018). *Intergovernmental panel on climate change*. Retrieved from www.ipcc.ch/

Jutila, S., Paloniemi, P., & Hakkarainen, M. (2017). *Proceedings of the heritage, tourism and hospitality international conference HTHIC 2017*, September 27–29, 2017 Pori, Finland. Turku: Turun yliopisto, pp. 109–113. Publication of Turku School of Economics, Pori Unit; nro A55.

Laine, M., Bamberg, J., & Jokinen, P. (2007). Tapaustutkimuksen käytäntö ja teoria. In M. Laine, J. Bamberg, & P. Jokinen (Eds.), *Tapaustutkimuksen taito*. (pp. 9–38). Helsinki, Finland: Gaudeamus.

Lambea Llop, N. (2017). A policy approach to the impact of tourist dwellings in condominiums and neighbourhoods in Barcelona. *Urban Research & Practice, 10*(1), 120–129. doi:10.1080/17535069.2017.1250522

Martin, C. J., Upham, P., & Klapper, R. (2017). Democratising platform governance in the sharing economy: An analytical framework and initial empirical insights. *Journal of Cleaner Production, 166*, 1395–1406.

Melé, D. (2003). The challenge of humanistic management. *Journal of Business Ethics, 44*, 77–88.

Middleton, V. T. C., & Hawkins, R. (1998). *Sustainable tourism: A marketing perspective*. Oxford: Butterworth—Heinemann.

Mihalic, T. (2000). Environmental management of a tourist destination: A factor of tourism competitiveness. *Tourism Management, 21*(1), 65–78.

Munar, A. M. (2011). Tourist-created content: Rethinking destination branding. *International Journal of Culture, Tourism and Hospitality Research, 5*(3), 291–305.

Munar, A. M. (2012). Social media strategies and destination management. *Scandinavian Journal of Hospitality and Tourism, 12*(2), 101–120.

Muñoz, P., & Cohen, B. (2017). Mapping out the sharing economy: A configurational approach to sharing business modeling. *Technological Forecasting & Social Change, 125*, 21–37.

Noirbnb. (2019). Retrieved from https://noirbnb.com/

Nuottila, J., Jutila, S., & Hakkarainen, M. (2017). Kirjallisuuskatsaus: Matkailun jakamistalous vastuullisuuden viitekehyksessä. *Matkailututkimus, 13*(1–2), 53–70.

Paloniemi, P., Jutila, S., & Hakkarainen, M. (2018). Paikallisten tarinoista ideoita matkailun kehittämiseen. *Matkailututkimus, 14*(2), 62–65.

Peeters, P., Gössling, S., Klijs, J., Milano, C., Novelli, M., Dijkmans, C., . . . Postma, A. (2018). *Overtourism: Impact and possible policy responses*. Brussels: Research for TRAN Committee European Parliament, Policy Department for Structural and Cohesion Policies. Retrieved from www.europarl.europa.eu/thinktank/en/document.html?reference=IPOL_STU(2018)629184

Pirson, M. (2017). A humanistic perspective for management theory: Protecting dignity and promoting well-being. *Journal of Business Ethics, 159*, 39–57.

Ritchie, J. R. B., & Crouch, G. I. (2003). *The competitive destination: A sustainable tourism perspective*. Oxford: CABI.

Saarinen, J. (2004). 'Destinations in change': The transformation process of tourist destinations. *Tourist Studies, 4*, 161–179.

Saarinen, J. (2010). The regional economics of tourism in northern Finland: The socio-economic implications of recent tourism development and future

possibilities for regional development. *Scandinavian Journal of Hospitality and Tourism*, *3*(2), 91–113.
Saarinen, J. (2014). Critical sustainability: Setting the limits to growth and responsibility in tourism. *Sustainability*, *6*(11), 1–17.
Sharpley, R. (2015). Tourism: A vehicle for development. In R. Sharpley & D. J. Telfer (Eds.), *Tourism and development: Concepts and issues* (pp. 3–30). Bristol: Channel View Publications.
Sheth, J., Sethia, N. K., & Srinivas, S. (2011). Mindful consumption: A customer-centric approach to sustainability. *Journal of the Academy of Marketing Science*, *39*(1), 21–39.
Sigala, M. (2017). Collaborative commerce in tourism: Implications for research and industry. *Current Issues in Tourism*, *20*(4), 346–355.
Tasci, A. D. A. (2017). Consumer demand for sustainability benchmarks in tourism and hospitality. *Tourism Review of AIEST—International Association of Scientific Experts in Tourism*, *72*(4), 375–391.
Tervo-Kankare, K., Kaján, E., & Saarinen, J. (2018). Costs and benefits of environmental change: Tourism industry's responses in arctic Finland. *Tourism Geographies*, *20*(2), 202–223.
Toni, M., Renzi, M. F., & Mattia, G. (2018). Understanding the link between collaborative economy and sustainable behaviour: An empirical investigation. *Journal of Cleaner Production*, *172*, 4467–4477.
Trunfio, M., & Della Lucia, M. (2019a). Co-creating value in destination management levering on stakeholder engagement. *e-Review of Tourism Research*, *16*(2–3), 195–204.
Trunfio, M., & Della Lucia, M. (2019b). Engaging destination stakeholders in the digital era: The best practice of Italian regional DMOs. *Journal of Hospitality & Tourism Research*, *43*(3), 349–373. doi:10.1177/1096348018807293
Tussyadiah, I. P., & Pesonen, J. (2018). Drivers and barriers of peer-to-peer accommodation stay—An exploratory study with American and Finnish travellers. *Current Issues in Tourism*, *21*(6), 703–720.
Visit Finland. (2019). Retrieved from www.visitfinland.com/sustainable-finland/
Volgger, M., Pforr, C., Stawinoga, A. E., Taplin, R., & Matthews, S. (2018). Who adopts the Airbnb innovation? An analysis of international visitors to Western Australia. *Tourism Recreation Research*, *43*(3), 305–320.
WCED. (1987). *Our common future. World commission on environmental and development*. Oxford: Oxford University Press.
Williams, C. C., & Horodnic, I. A. (2017). Regulating the sharing economy to prevent the growth of the informal sector in the hospitality industry. *International Journal of Contemporary Hospitality Management*, *29*(9), 2261–2278.
Yin, R. K. (2014). *Case study research: Design and methods* (5th ed.). Thousand Oaks, CA: Sage.

Part II

New Business Models for Creating Shared Value

5 Application of Slow Life Coaching Into an Agritourism Business Model

Daria Hołodnik and Kazimierz Perechuda

1. Introduction

The traditional view of agritourism has always been intrinsically linked with rural tourism in the way services are offered and delivered (mainly accommodation and food).

Today, however, staying at a farm could mean much more.[1] A wide range of tourism and leisure activities (e.g. outdoor activities, event organization, local and organic food, etc.) can be provided at the same level of professionalism and competitiveness as those provided by city hotels or hospitality suppliers. But the competitive edge of the first one can be delivered through an inspiring and natural landscape, a possibility to participate in folk art and culture, and a greater calmness as opposed to city surroundings.

From the tourist demand point of view, an ever-rising interest has been observed in tourists' choosing those hospitality objects that offer a slow life experience and slow food philosophy as well as eco-friendly services. Slow rhythms, a desire to explore traditions, a preference for having authentic experiences, and environmental sustainability are rarely realized in city tourism and hardly met in daily life; however, they are easily experienced during the 'suspended' time of a farm holiday (Sidali, Spiller & Schulze, 2011).

Since the traditional business model of agritourism services, though not effective because of low professionalism in the service delivery (in regard to those farms with a small interest in agritourism for a farm activities favoritism), has high potential for the slow life application, a new business model must be considered.

Thus, this chapter is mainly dedicated to discussing a process of designing competitive agritourism services aimed at solving humane problems through slow life teaching. In fact, two narration lines—transformation of the agritourism business model and slow life philosophy (as a means of achieving greater well-being and a harmonized feeling by agritourism guests)—are contrasted. Whereas the latter corresponds well with the humanistic management (HM) point of view, the former binds it with the service management (SM) field. When the aspect of service design is adapted to resolving various contemporary human problems, it can bring about great business success as well as help people find ways of achieving a more balanced life. This mixed approach has been used and emerged as a reflection about HM in the context of slow life application and service competitiveness; hence, it is discussed in the first section of the chapter.

In the following sections, a way to agritourism service transformation is explained by evaluating a case study research and analysis done at Lime Tree Valley Park, an agritourism object in Poland. Overall, the chapter has been divided into the following sections:

(1) Conceptual background of research problems (understanding humanistic management applied to the context of service management, and embedding agritourism services in the traditional and creative-tourism industries).
(2) Research problems and aims (investigation of two models of the agritourism value chain organization; hence, two ways of service management).
(3) Study design model and its research conduct (research methodology used in the case study investigation of Lime Tree Valley Park).
(4) Results presentation (managing agritourist experiences by means of slow life coaching at Lime Tree Valley Park).
(5) Conclusions (presenting two models of using agritourist slow life coaching—visible and invisible—as well as the synthetical summary of the chapter).

2. Conceptual Background of the Research Problems

The issue of incoherence of the humanistic concept in business is at the center of discussion in humanistic management theories and practices (Dierksmeier, 2016). However, the discrepancy between the humanistic and managerial understating of business may be reduced if the two sides of the matter are assimilated in a common approach, leading, so to speak, to the two co-creating the reality *per excellence* (integration of conceptualization and application). Undoubtedly, protection of intrinsic values and promotion of well-being is a base of the humanistic management identity (Pirson, 2017). Yet, embedding them in organizational contexts is essential to make them vivid in an interrelated socio-business life. So, if the humanistic point of view is obliged to stand for human values, but at the same time is immersed in the organizational background, then its validation depends on the right correspondence between these two. The authors have taken up an agritourism business model as the main context of cultivating humanistic values, which is understood as the applicable way of increasing service competitiveness (Figure 5.1).

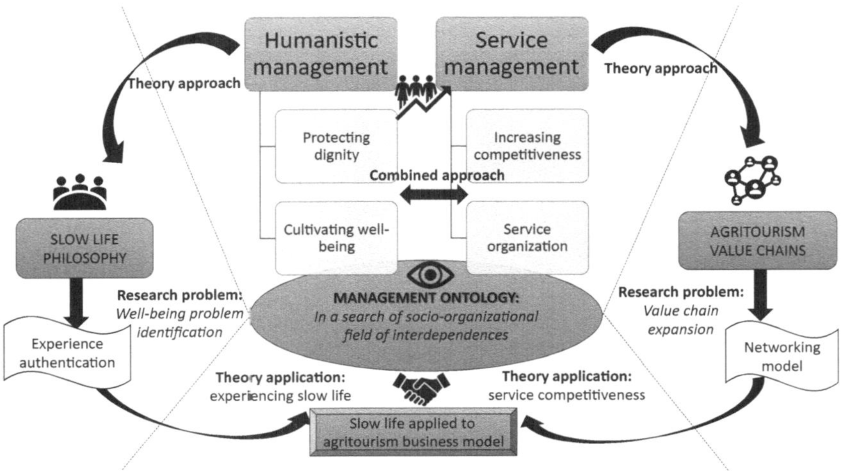

Figure 5.1 Application approach to humanistic management

Source: Authors' own elaboration

The fluid approach expresses the authors' intention to look for a synergy link between humanistic management (especially related to the aspect of well-being coaching) and service management (through value chains extension), as using them simultaneously enables us to construct a modern agritourism business model. So, it can be defined by two main characteristics: competitive, on the one hand, and oriented toward diminishing tourists' well-being problems, on the other. The key principle of making such a business model work is as follows: the more authentic a tourist's well-being problem is experienced, the more competitive are the services offered by agritourism. In the section about the case study conduct, the issue of how interviewed tourists became conscious of the 'empty' state of body and mind they previously had and how they were encouraged to make a stay more 'charging up' is presented.

Therefore, the theory approach made up from the combination of HM and SM outlines the extent to which slow life coaching could be helpful in both harmonizing the state of well-being of tourists and increasing agritourism service competitiveness. At the individual level, we have assumed that a slow life experience can result in a good educational effect, only if one were aware of its happening—in short, when tourists were actively taught how and why they should slow down the tempo of consuming tourism services or activities as well as keep a more conscious and respectful attitude whenever they do it. As for agritourism owners or managers, it is useful to acknowledge a value creation process in the course of changing a business model.

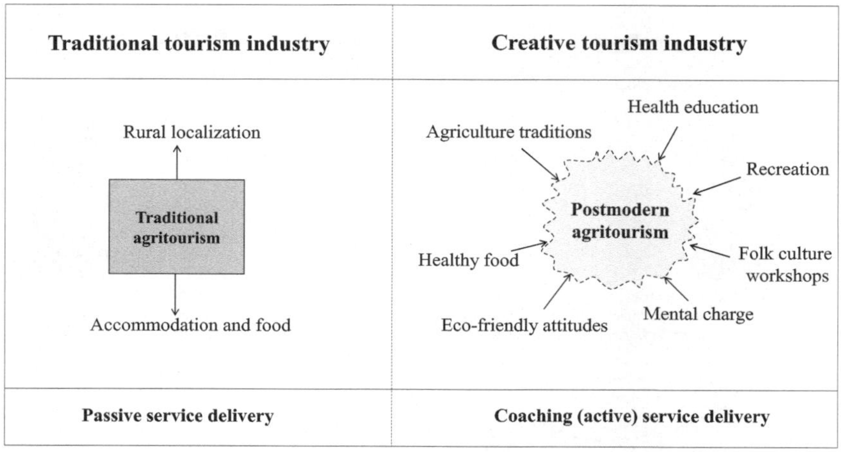

Figure 5.2 Agritourism business model in the traditional and creative industries

Source: Authors' own elaboration

Generally speaking, service competitiveness can be recognized by a value creation process within service delivery (Rong-Da Liang, 2017). In the agritourism field especially, it reflects the flexibility, creativity, and extension of service design (Broccardo, Culasso & Truant, 2017). Not every agritourism farm, even after having been designated as a 'slow life object', possesses the same value creation qualities; consequently, the authors have proposed distinguishing the agritourism service qualities attributed to traditional and creative industries (Figure 5.2). Staying at farm can be always regarded as a nice, relaxing, and calming experience just by the fact of spending time surrounded by nature. However, in reality, it is the range of offered services (active time spending) and a way they are delivered (genuine experiences) that differentiates the quality of creative agritourism' service from the traditional.

One of the vital prerequisites of designing a modern agritourism business model is a shift from the passive service delivery (a lack of time arrangement) to an active, educational service model (Hołodnik, 2017). It means that agritourists are offered not only food and accommodation but also, and above all, they can experience a certain lifestyle.

By rule, it is only when a farm with a business model of agritourism is able to generate meaningful, inspiring, and harmonious experiences for tourists that it be classified as a creative industry (Tomaszewska, 2018). In our approach, however, we have gone a step further by saying that these experiences are valuable and precious only when they are inspiring enough to make tourists realize how to become more and more conscious of the need to develop a slow lifestyle not only during a stay but also afterward as well. So, in our approach, we consider an authentic slow life experience as the starting point of teaching consumers, including tourists, how to live harmoniously, reduce tension and conflicts, and get rid of difficult emotions. So, in reality the difference between traditional and creative agritourism lies in the type of knowledge (slow life) and knowledge diffusion (authentic experiences), instead of an offered service itself (Hołodnik & Perechuda, 2014).

3. Research Problems and Aims (Two Models of the Agritourism Value Chains)

From the previous discussion, we discussed that a successful service delivery in agritourism relies on experience authentication. This standpoint is attributed to the humanistic management point of view, compared with the service management perspective which has a different understanding of what consumer experience management really is. For the latter, the main point is not to disappoint tourists with expected service delivery; in other words, respond as closely as possible to their preferences (Smith & Wheeler, 2002). This model, however, is based on delivering service satisfaction along with consumer education, because simply treating tourists as 'lords' does not teach them anything, nor make them more reflective. So, when consumer experience management is put into the context of humanistic understanding and agritourism adaptation, it requires a re-interpretation of the elements that make up service satisfaction (Figure 5.3).

Generally, the classical, linear process of service satisfaction, also referred to as 'customer value creation process' in the literature, is composed of two phases: service offer (equivalent to value identification) and service consumption (equivalent to value delivery). But in the agritourism business, we want to re-design them through the integration of two types of consumer experience management: consumer coaching (the humanistic approach) and satisfactory service delivery (the managerial approach).

Therefore, on the right side of Figure 5.3, the alternative model of a value chain oriented on slow life coaching and individual value creation is presented. However, what we have assumed from our former research[2] is that, to apply them successfully to the agritourism business, value chains have to be re-organized in the following ways: *from satisfaction delivery to experience authentication* (passive service offering is replaced by active identification of tourist problems) and *from service consumption to slow life coaching* (instead of consuming process, slow life solutions are provided by a network of coaches).

First, capability of slow life coaching mostly depends on establishing a humanistic relationship between an agritourist and an agritourism family (Choo & Petrick, 2014). These relationships are usually very warm,

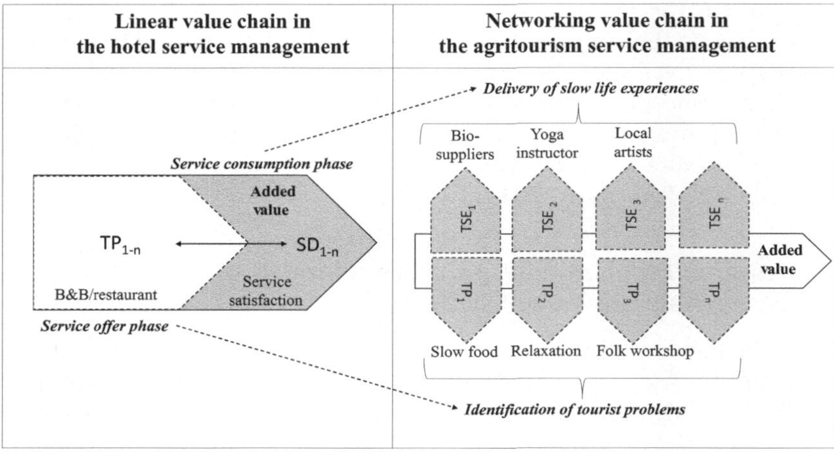

Figure 5.3 Linear and networking value chain in the agritourism service delivery

Legend: TP_1: tourist preferences (requested service); $SD_{1\text{-}n}$: service delivery (satisfactory if relevant to a request); $TP_{1\text{-}n}$: tourist problems; $TSE_{1\text{-}n}$: tourist slow life experiences (satisfactory if value transfer has worked)

Source: Authors' own elaboration

familiar, and friendly, so it has the potential to develop into a trustful relationship that functions as coach-coachee. Second, if a mutual friendship exists, it enhances generating the authentic experience for both the agritourist and the agricrew. For the guest (agritourist), it is very helpful when an agricrew can understand their way of thinking, living, and communicating. For the crew, what is helpful is reducing the chances of misunderstanding the intentions of the guest and getting rid of the hierarchical model of serving (i.e. instead of a 'servant-lord' relationship, a partnership co-creation is cultivated).

Third, to come up with a wider range of tourist experiences, the agritourism value chain must be networked with local specialists and suppliers (Renting, Marsden & Banks, 2003) that fit well to the slow life philosophy. So, once the tourist's well-being problem is identified, it should be matched with adequate solutions co-created by agritourism in collaboration with local coaches who are able to provide qualified slow life coaching in a particular service field (e.g. a local bio-supplier to offer homemade food, a yoga specialist to have the relaxation session, a local artist to organize the folk workshops, etc.).

To research what type of slow life coaching and agritourism service composition could be applied to resolve a specific group of well-being problems, see Table 5.1.

Table 5.1 Tourist problem identification: slow life applied to agritourism value chain

Group of well-being problems	*Type of slow life coaching*	*Service design capable of resolving problems*
Overstress, inability to keep work-life balance	**Health coaching:** • Relaxation and yoga coaching • Engaging in calming activities • Learning to 'do nothing' • No access to the Internet and other networks • Connecting with nature and contemplation	**Relaxing activities:** • Non-restricted limits of check-in, breakfast, and dining time • Deep sleep in closeness to nature and natural quiescence • Yoga sessions • Land walking • Going out for picnics • Reading in calmness
Lack of family, not enough time for spending with the family	**Social coaching:** • Family-like intimate climate of time spending • Person-to-person relationship building • Trust-building based on co-working	**Eco-activities:** • Active co-participation in farming lifestyle • Friendly atmosphere between guests and farm family • Home duties to fulfill, like helping in kitchen or learning or how to cook
Lack of connection with the nature and environment	**Ecology coaching:** • Fostering respect for the environment • Eco-friendly and smart systems of housekeeping • Paying attention to the use of daily resources • Re-connection with nature • Ecological coaching: open-air activities	**B&B:** • Using, when there is no need to change, the same towels for a couple of days • Eco-friendly heating systems • Reasonable water usage • Garbage segregation **Open-air activities:** • Active participation in garden caring • Digging in the field • Vegetable planting • Cow milking • Fruit harvesting
Lack of healthy food and diet	**Consumption coaching:** • Nutrition coaching • Cooking coaching • Discovering the joy of eating tasty and healthy foods	**Food service:** • Daily organic and garden-origin breakfast • Dining at regular hours • Eating less but with more aesthetic-looking meals • High-quality dishes • Herbal infusions during day • Dishes rich in fruit and vegetables • Local cheeses, ham, and milk products • Local beverages • Local alcohol

Source: Authors' own elaboration

Coaching by way of modern agritourism management means that the company actively assists its agritourists to find solutions to their health, social, and psychological problems. But to achieve this, an agritourist is offered the opportunity to engage in a variety of activities in the course of their stay at the location (Radwańska, Dąbrowski & Sokół, 2019).

Here, diagnosis of the tourist's explicit (e.g. visit inquiry) or implicit (e.g. private, family, or social problem) motivation constitutes a clue in designing and delivering solutions. Basically, planning an agritourist's destination itinerary and active advisory in service design depends on its accuracy and distinction (Stickdorn & Zehrer, 2009).

When this philosophy is discussed in terms of city tourists, its meaning is changed a little bit, mainly indicating alternative tourism places (e.g. restaurants, hotels, coffee bar, and so forth), in which healthy and local products (from eco-producers) can be served and local elements of handicraft or building materials are used. All of these implications are reflected in agritourism services and activities, but their application varies (Doh, Park & Kim, 2017).

4. Methodology and Study Design Model

The methodology of the research was based on the interpretative paradigm, according to which the socio-organizational phenomena do not exist by themselves but emerge from an actor interpretation over experienced situations, meanings, or undertaken intentions or actions (Czarniawska, 2004). Each time a content of experience manifests in front of different circumstances, a sense-making process has also been changed (Kostera, 2015). Therefore, in our research, we have assumed that the perception of an agritourism experience has a fleeting, unrepetitive, and ephemeral nature (Bauman Z, Bauman I., Kociatkiewicz & Kostera, 2015), yet its appearance can be caught in the interpretative moments (during the time of ethnographical discovery of agritourism experiences). So, in fact, the whole research procedure was designed in the manner of organizational ethnography, for which participation and observation are the essential methods of phenomena exploration (Figure 5.4).

Although the informal (first research visit) and formal (second research visit) phases of the research conduct were put into sequential order, collecting information was always done simultaneously: through hidden participation (i.e. when a researcher identity is unknown for an event or experience participants) and/or through open participation (i.e. a researcher becomes a watchful observer of an agritourism experience). The mysterious customer (hidden observation) was, however, used as a leading method because it is especially geared toward the study of tourism, leisure, gastronomy, and hospitality (Goodson & Phillimore, 2004) or when social interactions are part of a subject matter (Lancaster, 2005).

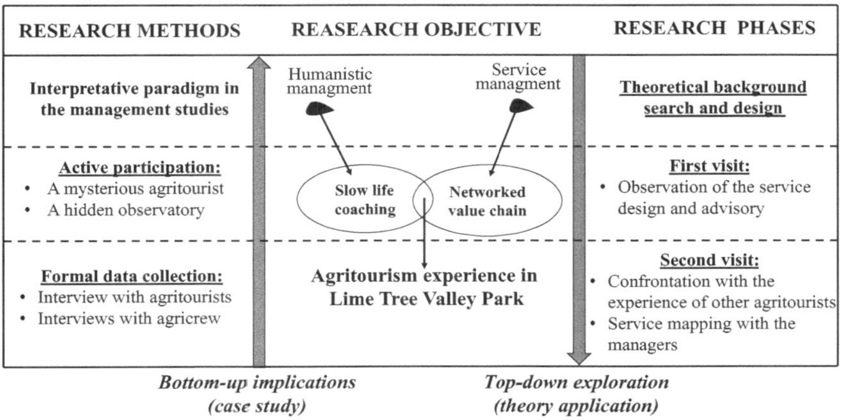

Figure 5.4 Study design model

Source: Authors' own elaboration

Both were the subjects of this research, since an agritourism experience mainly depends on service flexibility, openness, and personal design. The second reason for using it is that a shadow participation allows one to get an 'insider' understanding of how and why a particular interaction is interpreted in this or another way (Czarniawska, 2014).

In a case of agritourism, we looked at the slow life experience as a mutual interaction between an agritourist and agricrew and tried to understand both perspectives: a customer-coaching perspective validating the aspect of the slow life value transfer and a managerial perspective indicating the aspect of the service organization (see Figure 5.4.).

The main aim of the research was to identify the agritourist experience in the aspect of humanistic management and networked value chain. From the first point of view, the agritourist's services were observed in terms of high or low potential transfer of slow life values in:

- *Slow thinking* (cultivation of slow living and self-reflection),
- *Slow tourism* (high attention paid to choosing local-originated products, minimalization of unneeded consumption, high respect of culture, heritage, and local traditions),
- *Slow society* (cultivation of mutual respect, equality, openness, and kindness),
- *Slow agriculture* (wider sense of eco-agriculture and its importance to human health),
- *Slow environment* (sensibility for nature and active prevention against its damage).

Then, to get a closer picture of their implementation of the agritourism experience, interviews with agritourists and agricrew were conducted. Both sides were asked for:

- *Experience planning* (how important were slow life values in choosing services of this place: for the agritourist, or in providing services: for the agricrew?),
- *Experience design* (which of the slow life elements were engaged in service delivery?),
- *Experience delivery* (what attitudes/abilities of the slow lifestyle were trained in service delivery by agricrew or agritourist themselves?).

Furthermore, after one of the slow life experiences was revealed, an in-depth research was undertaken to check the level of impact it had on the agritourist's well-being problems, consciousness, and daily life education. This phase of the interview was aimed at evaluating the theoretical framework contained in Table 5.1, where potential well-being problems were matched with adequate agritourist coaching that presumably constituted a helpful antidote to resolve them. So, the agritourists indicated the type of slow life coaching that took place in their case and what was inspiring enough to be considered as a value transfer or authentic experience.

On the other hand, the means by which the agritourism value chain adopts or integrates the local sources and cooperation in order to organize a more attractive, authentic, and local transfer of values was verified. This perspective makes it possible to match agritourism experiences with the service networking that is a second theoretical grip attributed to service management. In other words, it measures the networking capacity of service design and customization (see Figure 5.4).

The central aim at this stage was to see how active and capable the agricrew was in advisory services and in cultivating a personal relationship between the guest and agritourist.

During the second research visit, staff members were asked for an interview in the form of open discussion and expert analysis. The questions were divided into the following sections:

(1) Linear value chain (key and supportive services offered by the agritourism itself)
(2) Networking value chain (cooperation with outsourcing service partners)

Generally speaking, identification of the value chain extension at Lime Tree Valley Park allowed us to analyze its overall potential for the agritourism business structure, particularly in the aspect of service design and coaching capability—consumer problems.

With regard to research objects, different business models of agritourism farms in Poland and abroad were selected; however, the analysis in this chapter is based on the case study of Lime Tree Valley Park, located in southeastern

Poland, not far from the town of Kazimierz Dolny.[3] This is an agritourism place that has succeeded in introducing slow life services and thus has been chosen as the benchmark of ecological and creative agritourism.

As for the methodological conclusions, it is worth pointing out that the quality of this ethnography research has mainly depended on a skillful, inspective observation and ability to be an inseparable part of the research field. These two make the interpretative paradigm and ethnography approach so difficult in practice, yet it can bring satisfactory results when there is a subtle balance between taking an active part and keeping a calm distance to experience reality.

5. Results

Lime Tree Valley Park is an extraordinary agritourism complex that offers five rooms, four apartments, and three cottages; local fine food; event organization for institutions and individuals; and many outdoor activities.[4] What was, however, at the heart of the research is the slow life character of their delivery that is tantamount to its successful application into the business model. As was discussed in the theoretical background, it is the consumer who participates wakefully in a provided service that can attest to the slow life coaching and experience. Therefore, the tangible elements of service delivery (such as a breathtaking localization, unique design, beautiful views, nice attractions, etc.) do not guarantee either correct identification of consumer problems or a way of slow life implementation. Figure 5.5 presents how the agritourism experience was co-created at Lime Tree Valley Park.

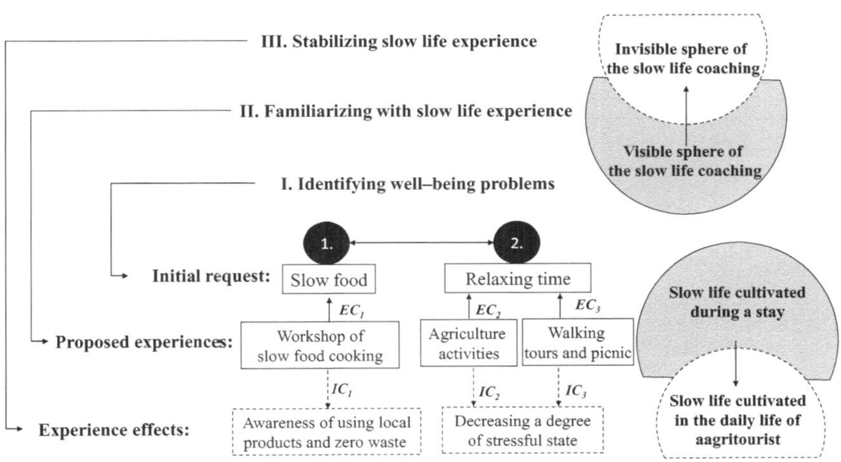

Figure 5.5 Development of agritourist coaching at Lime Tree Valley Park

Legend:

EC_{1-3}: explicit coaching, IC_{1-3}: implicit coaching

Source: Authors' own elaboration based on the research at Lime Tree Valley Park

The interviews with tourists were conducted with a couple staying at this agritourism resort for a weekend escape. Both of them agreed that the key reason for visiting Lime Tree Valley Park was to try local fine cuisine prepared homestyle and to rest in a calm-inducing natural environment (see Figure 5.5).

However, they did not plan any special itinerary for the stay. It was rather about having relaxation time and a restful weekend. But once they arrived and talked about plans for breakfast with the extremely friendly woman who was preparing the food, they began thinking about taking up some interesting activities. As the conversation continued, the couple was presented with the possibilities available at the site (activities at the farm) and around the area (activities outside). They then became interested in extending their stay (Table 5.2) for:

- Eco-agriculture activities,
- A healthy food workshop,
- A picnic and walking tours in the nearby forest.

It might be thought that construction of the agritourists' service design emerged from spontaneous talk with the farming woman, but this was not so. Observation and subsequent interviews with the manager led us to believe that the results were attributable to the highly educated, intelligent, and truly committed working staff co-creating the networking agritourism value chain models.

Most of all, the agritourists felt advised through slow cooking and active relaxation, additionally gaining more understanding of eco-agriculture

Table 5.2 Coaching design in value chain at Lime Tree Valley Park

Coaching effect	*Value chain composition*	*Service design*
Awareness of using local products and zero waste	**Networking value chain:** • Agricrew • Local farmers	**Workshop of slow food:** • Healthy food products • Recognition of locality • Homemade style of cooking • Cooking healthy and tasty meals
Decreasing degree of stressful state	**Networking value chain:** • Agricrew • Local forest guide	**Walking tours and picnic:** • Forest sessions • Forest education • Forest self-picnic
Decreasing degree of stressful state	**Networking value chain:** • Agricrew • Local farmers • Local community	**Agriculture activities:** • Cow milking • Animal feeding • Vegetable farming

Source: Authors' own elaboration based on study on the job at Lime Tree Valley Park

and local farms. They were very satisfied to learn how to live healthier and slower lives, even though it was only for a weekend stay. On being asked for the most valuable experiences, the participants identified the following:

- Co-emergence of leisure time creation,
- Speed in overcoming fatigue,
- Knowledge about eco-agriculture,
- Stress inoculation talks with the farming family.

They emphasized several times that all activities were arranged smoothly, naturally, and effortlessly, as if nobody had prompted them into doing anything. What could be done on site or nearby was simply very gently enhanced and introduced to them, but ultimately they had decided by themselves (see the self-realization model in Figure 5.6 later in the chapter).

So, there was no direct coaching; rather the procedure was converted into a subtle way of taking care of the agritourists' health and good time. One of the most appreciated pieces of experience was a kind of total freedom, a feeling or atmosphere of service process and not anything concrete (e.g. communication skills). But from the managerial perspective this is part of the consumer experience design and business mode in consequence. When managers of the analyzed agritourism were asked about any special strategy for arriving at the right agritourist advisory

Figure 5.6 Dualistic and self-approach coaching

Legend:

$PS_{1\text{-}n}$: Problem solving

$P_{1\text{-}n}$: Agritourist problems

$PR_{1\text{-}n}$: Problem realization

Source: Authors' own elaboration

and atmosphere (e.g. well-trained staff, system of communication skills, etc.), it turned out that that was not the case. But the agritourists' own words affirmed that the business model of Lime Tree Valley Park had been generally created according to the principle of 'the invisible coach'. To explain that idea, let's pass the narration to the manager:

> It is much better to assist agritourists select activities than to leave them make choices independently by themselves; the reason being that they are not familiar with the range of possible services offered and how the various events might fit their schedules. It is especially difficult to think and do anything when one is exhausted. We can understand that people need time to relax, and we don't want to put them through yet another stressful situation by requiring them to select services, make plans, and create structured days. We place high value on not coercing anybody to do anything. It must be the natural way of interacting, and the feelings are taken care of in addition to a friendly environment being created. Our agritourists should be given as much free space as they need it. The thing is to let them be inspired by themselves, while our role is finding their individual way in the perfect conditions for looking deeper and hearing louder.

The managers explained that the main source of this approach is related to slow life education within the agro-service design. As they pointed out, the difficult thing is to create an individual coaching framework when an agritourist's problems are unknown. Therefore, the network of slow life capabilities has to be extended to a rural destination where agritourism is inherently rooted (see Table 5.2, networking value chain). The leading and only strategy for the managers of an agritourism 'park' is to help agritourists resolve or be relieved of their contemporary problems by means of the slow life philosophy. However, its application to the service design is very difficult, as it depends on an agritourist's problem specification and individual preferences. Because some of the problems might be very inconvenient and uncomfortable to share, the managers have segmented them, depending on possible background, as follows:

- Physical reasons (e.g. muscle pain or body exhaustion),
- Psychological fears (e.g. lack of security or burnout by negative emotions),
- Social disharmony (e.g. overwhelming virtualization of a social life),
- Private life in decline (e.g. lack of family or of time to spend with them),
- Professional occupancy (e.g. no contact with nature and interest of environment).

Referring to the meaning of agritourist coaching stated by the manager, it is clearly connected with 'reconnecting of people with themselves by means of nature, happiness, and calmness'. These values have been successfully incorporated into Lime Tree Valley Park.

6. Conclusions and Discussion

Having in mind the fact that the coaching approach can be based either on external support, with a 'visible coach' (e.g. as is present in psychological therapy, medical treatments, fitness training, social media contacts, etc.), or self-elimination of problems (e.g. as is mainly used in natural therapies, reflexology, slow life, life awareness, etc.), the capabilities of the business model of Lime Tree Valley Park definitely refer to the latter (Figure 5.6).

The first approach is based on the traditional way of coaching, wherein a coach guides a guest participant by means of direct instructions. This approach can be called the 'dualistic approach', meaning that somebody from the agricrew is responsible for sorting out the problems of the guest. In this model of coaching, the real ability to diagnose the problems of the guest depends on the coach's skills and competences. Illustrating this in one's agritourist farm visit, it would be as though a kind of 'concierge' service was done in every zone of activities. It is not that the managers of Lime Tree Valley Park want to teach the agritourist but rather to enable them to self-manage their problems by embracing, even if just for a short time, a slow lifestyle.

Therefore, the main goal of invisible coaching is to create opportunities for the tourists to engage in a self-realizing and self-solving process. By itself, a farm stay naturally possesses such a quality. Most of the problems will disappear by themselves by simply creating for the guests the perfect space or conditions to find the time to relax and engage in enjoyable activities. This is the only principle in non-dual, space-based coaching (Dolot, 2018).

In conclusion, a slow life program in agritourism is aimed at coaching people on how to reinforce the body-mind energy in different areas of life (e.g. family, health, and environment), so the whole infrastructure, services, and customer relationship management have to be manifested in accordance with it. There is no unique framework for the coaching business model of the agritourism farm. However, for the process of examining a design of agritourist experience, the following analyses should include:

(1) Diagnosing the agritourist's time-spending concept (defining the main agritourist field of interests for each stay separately).
(2) Diagnosing the agritourist's current problems (matching with the relevant set of coaching activities).
(3) Designing daily activities (arranging for agreed agritourism services and local attractions).
(4) Co-creating agritourism space (working rhythm and conversation style).

Moreover, education of slow life through creating an authentic experience of calm, balanced, and harmonious existence is the way to apply the understanding of humanistic management into modern agritourism services. One of the reasons is that creative agritourism has enough qualities to teach postmodern people how to deal with a rushed, unbalanced, and disharmonized lifestyle that leads to nervous interactions with others, a feeling of being overwhelmed, and generally stressful and negative emotions. The organization of value chains managed by customer experience should be capable of generating meaningful moments, extraordinary impressions, valuable knowledge, and meaningful relations (Smith & Wheeler, 2002). In agritourism, this process refers to the arrangement of attractive time spending at the location delivered by means of slow life services.

The aim of the research, which constituted the focus of the chapter, was to analyze two aspects of the agritourism business model: slow life coaching and its application to the value chain model (agritourism service perspective). The case study of Lime Tree Valley Park in Poland shows how the former perspective can interfere with the latter. However, through different means of application, it is possible to attain a highly 'slow' and humanistic, yet efficient, service management. Here, it has been explained that application of an alternative (invisible) style of coaching is the most adaptable one for agritourism. Other conclusions indicate that:

(1) Orientation toward agritourist experiences is an indispensable trend to be investigated in agritourism management and marketing research.
(2) From the modern management perspective, service networking and co-sharing are the most applicable in agritourism.
(3) Agritourism services have much more potential if their value chains are enhanced by agriculture activities co-created by local and social capital.
(4) Slow life coaching put into the context of agritourism is the means to educate people on how to unify themselves within the existence of nature and locality.
(5) To distinguish itself from conventional hotels, an agritourism farm has to be able to create non-imitable core competences such as:

    - rustically preserved and decorated buildings,
    - Technical house equipped with agricultural tools,
    - Medium or small agricultural landscape,
    - Accessibility to panoramic views of the natural landscape,
    - No rush and slow lifestyle in provision of services,
    - Direct relationship between the agritourist (treated as part of the family) and the agricrew,
    - Field work and home duties as the recreational activities,
    - Flexibility of the services delivering,
    - Individual arrangements for a time-spending plan.

Notes

1. In some countries, agritourism farms are being recognized as exclusive objects, offering an opportunity to visit places with rare historical architectural designs and relive historical events. It especially regards locations that formerly served as the summer house (or villa) for the aristocratic and prominent families in a given region. Most of these types of places are located in Italy (e.g. *villa* or *castello* in Tuscany) or France (e.g. *chateau* throughout the country). In post-Soviet countries such as Poland, the Czech Republic, Slovakia, or Hungary, many castles in the rural areas have been renovated and now offer a very romantic, rustic, and charming stay (service comparable to the boutique hotel).
2. In the frame of a doctoral project and thesis at the Faculty of Informatics, Management, and Financing at Wroclaw University of Economics, 20 case studies from Poland, combined with 40 European wine agritourism cases from France, Italy, Austria, Germany, Hungary, and the Czech Republic, were analyzed.
3. The town of Kazimierz Dolny is situated in Lublin Province (Poland) on the right bank of the river Vistula. It is known for the unique Renesas center, art festivals, and attractive landscape. More information about the cultural heritage and leisure can be found at: www.kazimierz-dolny.pl/en/city/about-kazimierz-dolny/cultural-heritage
4. More information and a description of the services (accommodation, food, and events) offered by Lime Tree Valley Park can be found at: www.lipowadolina.pl/index.php

References

Bauman, Z., Bauman, I., Kociatkiewicz, J., & Kostera, M. (2015). *Management in a liquid modern world*. London: Polity.

Broccardo, L., Culasso, F., & Truant, E. (2017). Unlocking value creation using an agritourism business model. *Sustainability*, *9*(9), 1618. doi.org/10.3390/su9091618.

Choo, H., & Petrick, J. F. (2014). Social interactions and intentions to revisit for agritourism service encounters. *Tourism Management*, *40*(C), 372–381.

Czarniawska, B. (2004). *Introducing qualitative methods: Narratives in social science research*. London: Sage.

Czarniawska, B. (2014). Why I think shadowing is the best field technique in management and organization studies. *Qualitative Research in Organizations and Management*, *9*(1), 90–93.

Dierksmeier, C. (2016). What is 'humanistic' about humanistic management? *Humanistic Management Journal*, *1*, 9–32.

Doh, K., Park, S., & Kim, D. Y. (2017). Antecedents and consequences of managerial behaviour in agritourism. *Tourism Management*, *61*(C), 511–522.

Dolot, A. (2018). Non-directive communication techniques in the coaching process. *International Journal of Contemporary Management*, *17*(3), 77–100.

Goodson, L., & Phillimore, J. (2004). *Qualitative research in tourism: Ontologies, epistemologies and methodologies*. London and New York: Routledge.

Hołodnik, D. (2017). *Business models of wine agritourism farms*. Warsaw: CeDeWu.

Hołodnik, D., & Perechuda, K. (2014). Value co-creation in agritourism farms. In A. Frączkiewicz-Wronka, J. Gołuchowski, A. Adamus-Matuszyńska (Eds.),

Zeszyty Naukowe Uniwersytetu Ekonomicznego w Katowicach: Public Relations. Doskonalenie Procesu Komunikowania w Przestrzeni Publicznej (pp. 225–232). Katowice: Publishing House of the University of Economics in Katowice.

Kostera, M. (2015). *Research methods in the humanistic management*. Warsaw: Sedno.

Lancaster, G. (2005). *Research methods in management*. London: Routledge.

Pirson, M. (2017). *Humanistic management. protecting dignity and promoting well-being*. New York: Cambridge University Press.

Radwańska, K., Dąbrowski, D., & Sokół, J. L. (2019). Barriers and factors supporting agritourism and eco-agritourism activities in the bug valley municipalities in the opinion of farm owners. *Acta Scientarium Polonorum. Oeconomia*, *1*, 63–70.

Renting, H., Marsden, T. K., & Banks, J. (2003). Understanding alternative food networks: Exploring the role of short food supply chains in rural development. *Environment and Planning A: Economy and Space*, *35*(3), 393–411.

Rong-Da Liang, A. (2017). Considering the role of agritourism co-creation from a service-dominant logic perspective. *Tourism Management*, *40*(C), 354–367.

Sidali, K. L., Spiller, A., & Schulze, B. (2011). *Food, agriculture and tourism: Linking local gastronomy and rural tourism: Interdisciplinary perspectives*. Berlin, Heidelberg: Springer Science & Business Media.

Smith, S., & Wheeler, J. (2002). *Managing the customer experience*. London: Prentice Hall.

Stickdorn, M., & Zehrer, A. (2009). Service design in tourism: Customer experience driven destination management. In S. Clatworthy, J. V. Nisula, & S. Holmlid (Eds.), *Proceedings of 1st service design and service innovation conference, ServDes.2009. Linköping electronic conference proceedings* (p. 59). Linköping, Sweden: Linköping University Electronic Press.

Tomaszewska, A. (2018). Capacity of the tourism and the role of the creative industries- perception of visitors, barriers and motivation. *European Journal of Service Management*, *27/2–3/2018*, 489–498.

6 Cultural Heritage Triggering Corporate Investments

"Heritage Grab" or Sustainable Development?

Cecilia Pasquinelli

1. Introduction

The concept of cultural heritage has been constantly changing and is still expanding. A difference between tangible (e.g. physical items) and intangible heritage (e.g. oral traditions, performing arts, and rituals; UNESCO, 1972, cited in O'Brien, O'Keefe, Jayawickrama & Jigyasu, 2015) has been increasingly accepted, which testifies to an increasingly sophisticated approach to heritage conceptualization (Vecco, 2010). Beyond the traditional conception of heritage as coinciding with the remains of past glories and aesthetic grandeur, heritage concerns those activities and practices that, constructed in the present, are 'somehow' linked to the past (Kamel-Ahmed, 2015). This "living heritage" notion (United Nations, 2013) builds the bridge connecting past, present, and future of humankind's path of progress (O'Brien et al., 2015).

The complexity of dealing with time is an intrinsic dimension of heritage (Hall, Baird, James & Ram, 2016). Heritage is continually in a process of construction (Kamel-Ahmed, 2015). Only recently has the international community acknowledged that the conservation of cultural heritage concerns "places where social and cultural factors have been and continue to be important . . . rather than . . . a series of monuments offering physical evidence of the past" (United Nations, 2013, p. 13). A "functional approach" to defining cultural heritage is needed. From this perspective, cultural heritage is a dynamic capacity to "enrich people's lives" by providing inspiration and a sense of place and community through a lived and contemporary experience (MacKee, Askland & Askew, 2014). This definition prevails over an understanding of heritage as "frozen" in time (Hall et al., 2016), preserved and maintained "as it is" for future generations' passive admiration. Relationship building, people's engagement, and lively evolution are increasingly part of the cultural heritage notion. Even innovation (Sakdiyakorn & Sivarak, 2016) is not the antithesis of heritage, as heritage follows an unavoidable path of change.

These premises are important to this chapter, proposing a humanistic management perspective for outlining and analytically framing the role of corporate investors in cultural heritage and their capacity to boost local sustainable development. Corporate investors are intended in this chapter as private actors engaged with an economic process of value creation that, meeting economic constraints, may adopt "innovative approaches to enhance and extract value from cultural heritage" (Della Lucia & Trunfio, 2018, p. 36). The chapter proposes an analytical framework to reflect on local sustainable development, which combines the humanistic management perspective with land-grab literature, building on an analogy between land and cultural heritage as local resources for development.

Such literature helps conceptualize the dilemma of the corporate investor concerning the opportunity to behave as either a *heritage grabber* or a *heritage enhancer*. Land-grabbing literature allows us not only to cast a critical light on the privatization and hybridization of cultural heritage but also to reflect on those conditions that potentially allow the corporate investment to enhance the cultural asset's value, benefiting both the investor and the local community. Accordingly, the chapter discusses three important aspects of corporate investment, namely the governance, the business model, and the access to the resource (De Schutter, 2011), and complements the outlined framework in light of the humanistic management perspective. Prioritizing human needs, motivations, and living conditions within a "community of persons" (Melè, 2003), the humanistic management perspective represents an answer to the limit of the market paradigm in local development. The resulting framework helps reflect on sustainability in cultural heritage exploitation by stressing the continuous responsibility and agency of the corporate investor in the pursuit of human dignity and well-being.

2. Setting the Scene

A small village in Tuscany, Italy, named Castelfalfi, is part of the municipality of Montaione. The whole municipality has about 3,680 inhabitants. It is located in the Tuscan countryside, 50 km from Florence. Fairly unknown as a tourism destination regionally, Montaione/Castelfalfi has historically drawn significant tourist flows relative to its population size. In 2005, Montaione hosted about 36,300 arrivals, achieving 53,800 visitors in 2018. The tourism model corresponds to "tourism in rural areas" rather than "rural tourism", coinciding with mass tourism characterized by visitors looking for holiday flats for relaxation, aesthetic value, and comfort, rather than rural culture and agriculture, with limited engagement with the place (Berti & Rovai, 2011). In 2007, the Germany-based TUI Group, one of the major players in the global tourism industry with 19.5 billion euros in turnover in 2018 and 70,000 employees worldwide, invested in the purchase of the whole abandoned *borgo* (village) of Castelfalfi. The investment project included a huge regeneration project (originally, about 290 million euros), affecting the landscape and the medieval fabric of the village by restoring *casali* (ancient countryside houses) and transforming them into luxury flats and a four-star hotel, restoring a castle to host restaurants and shops, constructing a new five-star hotel, improving an existing golf course, and recovering and revitalizing agricultural assets around the *borgo*.

Through a formal public debate steered by local government, a participatory process took place before the modification of the urban regulatory framework, which was necessary to allow TUI's project to be implemented. This public debate, which took place in 2007, is considered one of the best examples of citizen participation in planning (Zetti, 2010). During the public debate, the concerns of the local community emerged. These concerns focused on the size of the project, the new construction, the style of the restoration, the impact on the hilly landscape, the integration of Castelfalfi with the local system, the need for TUI to guarantee openness and access to the *borgo*, the shops and services within it, and the role of agriculture in the regeneration of the area. Although there is no analysis of the impact of TUI's investment in the territory and the evolution of the destination, TUI's Toscana Resort Castelfalfi seems to have contributed to the relaunch of a mature destination not only by diversifying local offerings, creating new jobs, and drawing international events such as the Peugeut Drone Film Festival in 2017 but also by hosting smaller and local events for visitors and residents and steering commercial activities that are open to the outside world.

Sammezzano Castle, located 50 km from Florence, Italy, is an example of eclectic architecture with oriental inspiration. In the 19th century, the marquess Ferdinando d'Aragona transformed the historical castle, which dates back to Romans, into a colourful building full of art pieces made of handicraft tiles and plaster and surrounded it with a park with the highest concentration of sequoia trees in Europe. Under the protection of the law since the 1920s for its historic, artistic, and natural prestige, the castle was sold to a company that obtained the right to transform the castle into a hotel and restaurant in 1970.[1] Since 2000, the castle has been abandoned. Although the castle is considered a cultural heritage asset, restoration and preservation interventions have not occurred; the castle remains closed and inaccessible, the target of vandalism and thefts. In 2017, an auction assigned the property of the castle to the Dubai-based company Helitrope Limited for 15.4 million euros, but the sale was later nullified. No investor showed up in subsequent auctions. In 2015, the citizens' movement Save Sammezzano started its activities to boost attention for the castle and its artistic and cultural relevance. Active on social media platforms and instigating concrete initiatives, Save Sammezzano promoted the candidacies of the castle to the Italian Environmental Fund's "Hearth Places" contest, which the castle won; to the 2018 World Monuments Watch; and to Europa Nostra's 7 Most Endangered sites listing, resulting in its being ranked among the 12 sites at greatest risk and most deserving of restoration.[2] However, the lack of a definite property owner is an obstacle to any restoration interventions.

Solomeo is a medieval village in the Umbria region, central Italy. The castle on the top of the hamlet, signaling the medieval roots of the village, was restored by Brunello Cucinelli whose company, which produces luxury cashmere garments, moved into in 1987. Brunello Cucinelli is one of the most iconic Made in Italy fashion brands, positioning in luxury segments with a net revenue of 607 million euros in 2019, recording a 9.9% increase over the previous year. In the 1980s, the village was losing much of its population; the entrepreneur understood that in the future people would again value villages, which are rich in history and culture, and come back to them, outside of large and crowded urban areas (Bianchi Martini, Corvino & Minoja, 2015). Over time the company expanded its presence in the village, and a new structure in the valley below the castle is currently occupied by the company headquarters and artisanal workshops where most products are manufactured. Today, the castle hosts the School of Arts and Crafts, set up by the entrepreneur in 2013, to strengthen the continuous learning and specialized manual skills utilized in high-quality manufacturing. Solomeo is branded as "the Hamlet of the Spirit" by Brunello Cucinelli, and it is the repository of the symbolic values the luxury brand stands for: dignity of human beings and beauty. The hamlet and the person brand of the entrepreneur (a "humanist-businessman", Picchio, 2019) make the company highly reputable in global luxury markets, physically embodying practices of production responding to a "gracious growth" principle and communicating business as a way to combine profit with ethical, moral principles and dignity (Brunello Cucinelli SpA, 2019). Working conditions, involvement of locals and local small businesses and significant investments in the hamlet continuously strengthen the relationship of the company with the village, which is a unique tourism destination among several medieval villages in the rural areas of Central Italy. The creation of a cultural system named "Foro delle Arti" (Forum of Arts) composed of a theatre, an amphitheater and the "philosophers' garden" where concerts and cultural events are organized (Bianchi Martini et al., 2015) contribute to local quality of life and make the hamlet an appealing destination as "pleasant periphery". Also, attracted by the reputation of Brunello Cucinelli and the resonance of the brand, visitors experience firsthand the entrepreneur's philosophy in Solomeo. A strong identification exists among the village, its community, and the entrepreneur: the hamlet is the laboratory and is meant to be the tangible evidence of a "humanistic capitalism" enacted by the entrepreneur and his brand.

3. Conceptualizing and Framing Emerging Issues

Different research streams help frame the issues emerging from the aforementioned cases. First, cultural heritage has been increasingly considered and discussed by policy-makers and scholars as a vital resource supporting local development and urban regeneration. The cultural turn in local development witnessed by the attention given to culture-led regeneration processes (Plaza, 2000; Bailey, Miles & Stark, 2004; O'Brian & Miles, 2010) and to the evolution of cultural economy (Scott, 1997) highlights the positive socio-economic role of culture in contemporary society. In this framework, tourism development has greatly contributed to the implementation of the cultural paradigm in urban planning and city management. In this regard, the coupling of cultural heritage and tourism has been largely discussed in relation to the activation of virtuous circuits of regional competitiveness boosted by their intertwining (Alberti & Giusti, 2012) and by the hybridization of local heritage and the creative economy (Della Lucia, 2015). The academic debate, however, also stressed that the use of cultural heritage as leverage for tourism development and urban regeneration may create negative issues of cultural commodification (Scott, 1997; Zukin, 1995), standardization, and "placelessness" (Evans, 2003), a controversial construction of authenticity according to tourists, residents, and governments (Cole, 2007), and threats to the conservation of material heritage under the pressure of unprecedented tourism growth (Garcìa-Hernandez, Calle-Vaquero & Yubero, 2017).

Second, the privatization and the hybridization of cultural heritage connected to the selling of state-owned historic real estates have been discussed (Benedikter, 2004; Ponzini, 2010), as have the commercial exploitation of local culture through films, creative productions (Power & Scott, 2004), events, and festivals (Jakob, 2012) and the increasing number of public-private partnerships for the promotion of cultural assets, legitimized by a collective interest in urban regeneration and local development. Some scholars maintained a neoliberal agenda in built heritage management (McGuigan, 2005; Negussie, 2006), a marketing-driven appropriation of intangible cultural heritage (Muresan & Smith, 1998; Starr, 2010), and processes of falsification of heritage to meet the market "momentum", not without conflicts between preservation and the building of a satisfactory tourist experience (Nuryanti, 1996).

However, cultural asset privatization and hybridization may infuse capital and entrepreneurial vitality into abandoned heritage sites, thus triggering tourism development and revitalization. Private actors can play a positive role in the cross-fertilization of cultural heritage and creativity, resulting in innovative ways of creating value by leveraging local culture and putting in place strategies to engage local stakeholders with regeneration processes (Della Lucia, Trunfio & Go, 2017; Della Lucia & Trunfio, 2018). Cultural entrepreneurship has, in fact, the distinctive ability "to draw on tangible and intangible cultural factors, to harness stakeholder collaboration to create value propositions which attract audiences" (Go, Lemmetyinen & Hakala, 2014, p. 6).

Two scenarios emerge from such discussion. The first relates to corporate investments in cultural heritage that are capable of triggering tourism development while supporting and strengthening the value of the cultural asset, not only by creating value for the tourism industry and guaranteeing the asset's physical existence in the long run (e.g. restoration) but also by contributing to making it a "living heritage" (United Nations, 2013). This is possible by steering and fostering an evolving relationship of the cultural asset with local community. In this scenario, the business investor becomes a "heritage enhancer". In this case, there is a tendency toward sustainable tourism that, based on the connections among the environment, local communities, and tourism (Liu, 2003), includes "cultural sustainability" as the forth pillar of sustainable development (Soini & Birkeland, 2014).

The second scenario concerns the business investor acting as a "heritage grabber", disconnecting the cultural asset from the local community for the exclusive sake of private interest and market utility. While assuring the physical existence of the cultural asset for the pursuit of tourism development and the related economic objectives, the investor overlooks and undermines the opportunity to maintain or potentiate the asset as "living heritage".

3.1 *Investing in Cultural Heritage: An Analogy With the Land-Grab Phenomenon*

In the light of two emerging scenarios (i.e. heritage grabber *versus* heritage enhancer), we refer to the land-grab literature to explain the modalities and dynamics according to which the corporate investment takes place. This consolidated literature supports the exploratory effort made in this chapter to frame and explain corporate investors' behaviors and attitudes to cultural heritage. The title of this chapter is inspired by a report by Cotula, Vermeulen, Leonard and Keeley (2009) titled "Land grab or development opportunity?" to stress how the analogy is utilized to frame the issues discussed in the chapter. The land-grab literature helps highlight the critical aspects to be considered when dealing with cultural heritage as "resource" for local development and suggests three analytical dimensions to understand corporate investors' approach to cultural heritage.

The term 'land grab' (Cotula et al., 2009; Zoomers, 2010; Borras & Franco, 2010) refers to large-scale (trans)national commercial land transactions (Borras, Hall, Scoones, White & Wolford, 2011), especially in the context of developing countries going through processes of change in land use and land property. The analyses of land grab so far produced in the literature make a two-fold phenomenon emerge. On one side, large-scale investments are said to weakly reduce poverty compared with the alternative of improving land and water supply to local farming communities; they increase the export of agricultural productions (Lavers, 2012), raising food security issues (Cotula et al., 2009), and even when an effort is made to protect land users from eviction, the destructive effects on local livelihoods increase with commercial pressure on the land (De Schutter, 2011).

On the other side, the literature witnesses that, especially when "land is underutilized and abundant", the benefits of large-scale investments are evident in terms of increases in the provision of goods and services, job creation, and access to technology and markets (World Bank, 2010, cited in De Schutter, 2011). This point makes the proposed analogy (i.e. cultural heritage-land) particularly meaningful in contexts having extensive endowment of cultural assets owned by the state and municipalities (Formato & Presenza, 2018) which struggle in deploying sufficient financial resources to guarantee their restoration and functioning. Following the Portuguese and Spanish experiences, the Italian government promoted a national program, "*Valore-Paese-Cammini e Percorsi*", that aims to sign concession agreements drawing long-term private investments in areas where regeneration and tourism-led development are pursued.

It is worth stressing the limitations of the proposed analogy. The investments in cultural heritage are different from land investments concerning industrial domains and geographical contexts. On the other hand, the motivations for the application of such an analogy concern the role of corporate investors with financial capacity to invest in "underutilized and abundant" resources; the impact of such investment on local communities and their path of development; and the development of local systems blocked by small-sized and fragmented entrepreneurial fabric.

Three analytical perspectives are proposed to frame the analogy and explain modalities and dynamics according to which corporate investments in cultural heritage assets take place. These are governance, business model, and access. Proposed by De Schutter (2011), these three dimensions are presented to entangle the corporate investment in cultural heritage.

The *governance* perspective refers to the capacity of different actors to play their roles and, particularly, their capacity to adopt optimal bargaining positions. Investments in developing countries that are often characterized by weak governance may carry significant risks for the region and local communities (Deininger, 2011). In relation to the focus of this chapter, this point draws attention to the maturity of the cultural heritage governance in order to evaluate the concrete risks of corporate investments. The governance perspective reflects upon the interactions among the different relevant players, namely the private actors (e.g. local actors and large-scale investors) and the public actors whose regulating role is deemed fundamental to setting the modalities of resource usage. The involvement and engagement of local communities in the process of regulating the presence of the investor in the region is a salient condition for the minimization of the risks connected to large-scale investments.

The *business model* locally adopted by the corporate investor is made possible by local circumstances and regulations. That is, the business model relates to the private investor, but it highly depends on what public regulations allow or forbid. If the governance perspective focuses on decision-making and the capacity of the different actors to raise their voices, the business model perspective refers to the capacity to design the functioning of an investment within a certain region, anticipating and shaping future interactions among all the actors. Through a bargaining phase of the investment, local government may play an active role in making the business model fit with local contexts. This, however, is not always the case in practice. Furthermore, local governments face difficulties in projecting the local systems and their needs in the future (put simply, what business model will suit a local community and its needs in the future?).

Access and, particularly, the definition of rules for protecting forms of access to and usage of a resource, beyond formal property rights, is extremely important to design the modalities in which the corporate investor will impact local community. Such rules are particularly relevant in the case of land that was originally communal, offering locals opportunities for fishing, hunting, and gathering. Access matters to cultural heritage, too. Heritage, in fact, "enrich[es] people's lives" and inspires a sense of place in local communities (MacKee et al., 2014).

3.2 From "Heritage Grabber" to "Heritage Enhancer": The Humanistic Management Perspective

The three dimensions emerging from the land-grab debate compose the framework for analyzing corporate investments in cultural heritage and their role in local sustainable development. However, they are not enough to entangle such a role in the practice. The three dimensions miss including a dynamic perspective that, instead, highly matters when considering long-term impacts. Participation and regulation attempting to design access and to influence business models are necessary yet not sufficient conditions. Limited forecasting ability, planning capacity, and unbalanced bargaining power may limit the explanatory power of the three described analytical dimensions (i.e. governance, business model, access). Provided that public intervention in regulating and monitoring the process of investment is key to making corporate investments in cultural heritage contribute to local sustainable development, another important dimension has to be added to the framework: the humanistic management perspective.

The term 'humanistic management' describes the approach to business management that prioritizes human needs, human motivations, and human conditions and is oriented to "the development of human virtue to its fullest extent" (Melè, 2003, p. 78). The humanistic approach may be an answer to the limits the market paradigm revealed during the recent financial crisis, urging us to rethink management approaches and their exclusive economic rationale (Melè, 2009). It may also be an answer to the limits of coercive power and regulations in shaping sustainable development, as stated previously.

The inter-organizational model of humanistic management suggests that "humanism" implies looking beyond the borders of a single organization (Rocha & Miles, 2009), thus integrating an interest for the individuals who live in the surrounding environment into the central notion of a "community of persons" (Melè, 2003). Accordingly, if traditional literature has sustained the role of the external environment in shaping business competitiveness, humanistic management extends business responsibility and concern for human dignity to the territory hosting the investment and its inhabitants. This characteristic stresses the relevance of humanistic management to cultural heritage investments in a given territory and, so, to the local community.

In light of these premises, Spitzeck's (2011) integrated model of humanistic management is utilized to frame business investments in cultural heritage. Using profit-making and humanistic rationale as dimensions of the proposed model, this author identified three typologies of business conduct that are useful for disentangling the role of the corporate investor as either "heritage grabber" or "heritage enhancer". Such business conducts are described as follows: 1) challenge for the organization, 2) challenge for society (heritage grabber), and 3) humanistic management (heritage enhancer).

These three business conducts are described next and followed by interpretative hypotheses referring to the cases introduced in section 2. The cases help show the application of the proposed framework through the formulation of working hypotheses that empirical research should test.

Challenge for the organization: Unlikely private investment. This typology refers to potential investments in cultural heritage that are not profitable for various reasons, such as location and usage constraints that undermine the appeal of the cultural asset in the investment market. Because investment in cultural assets often requires significant financial resources, profit and returns on investment need to be guaranteed. A lack of such guarantee is a challenge for the organization, which may decide not to invest. Business investments are here considered to be radically different from patronage, which instead may occur in contexts with no prospective profits.

In the Sammezzano case, the hypothesis of an unlikely corporate investment follows the evident difficulties in the sale of the castle, as emerged from the collected information. The long-term legal controversies and the uncertainties surrounding the castle may represent an obstacle and a cost for potential investors, to be added to the massive financial investment required to guarantee high standards of preservation and restoration. In a context of significant social mobilization and rising media coverage transforming the castle into a recognizable landmark with growing symbolic value, investments in Sammezzano have not concretized yet.

Challenge for society: Likely heritage grabber. This typology describes profitable investments in cultural heritage by business investors exclusively oriented to the market and immediate economic utility. The private investor of this type tends to seek investment projects in contexts allowing for simplified and rapid decision-making. This may be the case of contexts in which *governance* is restricted to key players (the seller and the buyer) of a commercial negotiation. The *business model* is entirely up to the private investor, who does not have limitations such as specific functions and activities to carry out in the cultural site, architectural constraints related to the size or quality of interventions, or requirements such as job creation or restoration or regeneration goals. The larger the governance, the more numerous the requirements and constraints. According to the business model, exclusive *access* to the resource (i.e. cultural asset) can be granted to the clients of the investing firm, while no access is guaranteed to the public audience. This exclusivity of access transforms the cultural asset into a closed and self-contained physical and functional structure, like traditional resorts. In this case, any intervention in favor of the local community's access and active usage of the cultural asset is possible and derives from a deliberate and voluntary act of philanthropy of the company's management.

In Castelfalfi, a risk of heritage grab was perceived by at least part of the local community during the first phase of the investment project, as the documents related to the public debate suggest. Even though there is no evidence to state the company adopted this approach, the reaction of the local community exemplifies this second typology. The public debate, in fact, raised concerns about the size of the project, the impact on the landscape, a style too unlike local culture, and the need for integration with the local system. Furthermore, guarantees of openness and access to the *borgo* were asked, and the need to significantly integrate agriculture in the ambitious regeneration plan was affirmed.

Humanistic management: Heritage enhancer. This typology posits that forms of humanistic management are necessary for the private investor to function as a heritage enhancer. In line with Spitzeck's thesis (2011), we distinguish two modalities, as follows:

1) "Instrumentalising business": This firm is aware that, without heritage preservation in pursuit of cultural sustainability, the cultural

> asset will lose its value, thus undermining the profit-making potential of the investment. In this case, profit drives the investor's humanistic attitude. The private investor seeks contexts in which decision-making is carried out by *governance* configurations embracing a multitude of actors. These actors contribute to the project and will be "on board" during the implementation phase, with low degrees of conflict. Various actors' involvement and the local community's engagement impart authenticity to the cultural asset and, thus, value to the business. The *business model* is up to the private investor, but rules and requirements emerge from the participative decision-making. *Access* is guaranteed to the public audience, and locals have free access to the heritage site. The presence of and relationship to the local community are key sources of value creation for the firm in the tourism industry.

This way of conceiving of humanistic management may evolve problematically. As Spitzeck (2011) suggested, if profit motivation decreases over time for the instrumentalizing business, then there is a risk of a change in the managerial approach such that the investor may shift from being a heritage enhancer to a heritage grabber. We must consider the additional monetary costs the heritage enhancer sustains to shape forms of open access and "humanistic" business models. Initial rules and requirements may change or weaken over time, for instance, under the threat of divestment that, once the investment is completed, creates serious socio-economic problems for the local community, such as an increase in unemployment, as well as political consensus issues. At that point, the preservation and enhancement of the heritage asset's cultural value are in the hands of the investor, whose priorities, as said, may change.

Castelfalfi is hypothesized as a potential case of heritage enhancement based on the "instrumentalising business" approach. The governance dimension played a role in this case: the investor entered in a tight dialogue with the local community and government, as the public debate process and related documents revealed.

In the implementation phase, events and commercial activities in Castelfalfi and the capacity to draw local presence in the *borgo* are valuable assets not only for the community but also for the company, raising the authenticity of the Castelfalfi experience in the Tuscan hills. As the investor replied to the requests that emerged from the public debate, the targeted segments are cultural and wellness tourists seeking events and cultural activities, potentially interested in local products and appealing experiences. Considering the relatively recent investment (the last opening in the *borgo* was the five-star hotel in 2017), the proposed theoretical framework suggests the relevance of adopting an evolutionary perspective on the "instrumentalising business". An analysis of future developments is necessary to understand continuity and strength of effort in steering and favoring the relation of the local community with the *borgo*.

2) "Truly humanistic management": Private investments in cultural heritage guarantee cultural sustainability and sustainable development in the long run. The pursuit of "unconditional human dignity" (Melè, 2003) and the legitimate and responsible corporate conduct pursuing the vision of serving society and fostering citizens' quality of life are the pillars of humanistic management. Profit-making is not the only priority, although it is fundamental to business (otherwise the organization will face a challenge, according to Spitzteck). Humanistic motivations are the driver of business investments. This option is the necessary condition for turning the principle of sustainability into concrete and durable action.

The case of Solomeo is hypothesized to fit with the "truly humanistic management" modality, given the available information and the longstanding relation the company has established with the village and the local community, until identifying the global fashion brand with it. The humanistic approach emerges from the commitment to those values that are a common thread across manufacturing, human resource management, and design of space, including public space. It is worth noting that humanistic management in this case, also thanks to the reputation of Brunello Cucinelli's brand, becomes the motivation and key attraction for tourists, willing to see and experience the laboratory of the entrepreneur's philosophy, which—we can argue—is hybridizing local cultural heritage.

The proposed framework to analyze corporate investments in cultural heritage and their capacity to boost local sustainable development is summarized in Table 6.1, by taking into account the three dimensions

Table 6.1 Hypotheses for future research based on the proposed framework

	Challenge for the organization: Unlikely private investment	*Challenge for society: Likely heritage grabber*	*Humanistic management: Heritage enhancer*	
			Instrumentalizing business	*Truly humanistic management*
Governance Business model Access	Sammezzano?	Castelfalfi (I phase)?	Castelfalfi (II phase)?	Solomeo?

Source: Author's own elaboration

emerged from the land-grab literature (De Schutter, 2011) and humanistic management (Spitzeck, 2011). Working hypotheses are formulated about the cases, as explained earlier, opening to future empirical research.

4. Conclusion

This chapter proposed a humanistic management perspective to interpret the role of the business investor in cultural heritage and integrated it with land-grab literature to provide an analytical framework. The framework is meant to support future empirical analyses aimed to assess the capacity of corporate investments to boost local sustainable development. Further theoretical insights can derive from the application of this framework, until suggesting the conditions under which the impact of such investments can be positive in the long run.

Three cases—Castelfalfi and Sammezzano in Tuscany, and Solomeo in Umbria, Italy—were used to provide concrete background for the proposed conceptualization. The combination of different streams of literature concerning the role of cultural heritage in tourism development and the implications of the increasing role of private investments in cultural assets resulted in the definition of a dilemma for the corporate investor: deciding to function as either a heritage grabber or a heritage enhancer. Borrowing analytical perspectives from land-grab literature, three dimensions for distinguishing heritage grabber from heritage enhancer were utilized: governance, business model, and access. The relevance of these dimensions emerged in the three cases: in Castelfalfi, the governance dimension prevails in explaining the heritage enhancement. Access is central to the three cases: in Sammezzano, grassroots mobilization and the volunteers organizing forms of temporary access to the castle for visitors stress that, beyond the physical abandonment, this cultural asset continues to be "living" and socially and culturally relevant to the local community. The salience of access is confirmed by the results of the public debate in Castelfalfi since one of the key points was the explicit request of guarantee of access to the *borgo*. Finally, the company's business model is the prevailing aspect in the Solomeo case: especially, the identification of the hamlet (and the people living and working in it) with the company's core values is at the core of Brunello Cucinelli's brand.

Humanistic management, completing the framework, contributes to explaining the heritage grabber's and heritage enhancer's conducts, motivations, and behaviors by stressing the continuous responsibility and agency of corporate investors in the pursuit of human dignity and well-being.

This framework opens to future research. The proposed hypotheses should be tested through dedicated empirical efforts. The aforementioned cases were taken from the Italian context, belonging to peripheral and rural areas in Central Italy, with similar cultural and socio-economic characteristics. Other geographical contexts should be considered to validate the proposed framework, to see if and to what extent the role and relevance of the different framework components change when explaining heritage grab and heritage enhancement from a sustainable development perspective.

Notes

1. Information retrieved from www.savesammezzano.com, last access 11 July 2019.
2. Information retrieved from www.savesammezzano.com, last access 11 July 2019.

References

Alberti, F., & Giusti, J. (2012). Cultural heritage, tourism and regional competitiveness: The Motor Valley cluster. *City, Culture and Society*, *3*, 261–273.

Bailey, C., Miles, S., & Stark, P. (2004). Culture-led urban regeneration and the revitalisation of identities in Newcastle, Gatehead and the north east of England. *International Journal of Cultural Policy*, *10*(1), 46–65.

Benedikter, R. (2004). Privatisation of Italian cultural heritage. *International Journal of Heritage Studies*, *10*(4), 369–389.

Berti, G., & Rovai, M. (2011). Turismo rurale o turismo incorniciato nel paesaggio rurale? Il caso di Montaione. In A. Pacciani (Ed.), *Aree rurali e configurazioni turistiche. Differenziazione di sentieri di sviluppo in Toscana* (pp. 153–216). Rome: Franco Angeli.

Bianchi Martini, S., Corvino, A., & Minoja, M. (2015). Brunello Cucinelli. In M. Minoja (Ed.), *Bene commune e comportamenti responsabili*. Milan: Egea.

Borras, S., & Franco, J. (2010). Towards a broader view of the politics of global land grab: Rethinking land issues, reframing resistance. *ICAS Working Paper Series*, *001*, 1–39.

Borras, S., Hall, R., Scoones, I., White, B., & Wolford, W. (2011). Towards a better understanding of global land grabbing: An editorial introduction. *The Journal of Peasant Studies*, *38*(2), 209–216.

Brunello Cucinelli SpA. (2019). Brunello Cucinelli: The board of directors has examined the 2019 preliminary results. Retrieved February 21, 2020, from www.brunellocucinelli.com

Cole, S. (2007). Beyond authenticity and commodification. *Annals of Tourism Research*, *34*(4), 943–960.

Cotula, L., Vermeulen, S., Leonard, R., & Keeley, J. (2009). *Land grab or development opportunity? Agricultural investment and international land deals in Africa*. London: IIED/FAO/IFAD.

De Schutter, O. (2011). How not to think of land-grabbing: Three critiques of large-scale investments in farmland. *The Journal of Peasant Studies*, *38*(2), 249–279.

Deininger, K. (2011). Challenges posed by the new wave of farmland investment. *The Journal of Peasant Studies*, *38*(2), 217–247.

Della Lucia, M. (2015). Creative cities: Urban experimental labs. *International Journal of Management Cases*, *17*(4), 156–172.

Della Lucia, M., & Trunfio, M. (2018). The role of the private actor in cultural regeneration: Hybridizing cultural heritage with creativity in the city. *Cities*, *82*, 35–44.

Della Lucia, M., Trunfio, M., & Go, F. M. (2017). Heritage and urban regeneration: Towards creative tourism. In N. Bellini & C. Pasquinelli (Eds.), *Tourism in the city. Towards an integrative agenda on urban tourism* (pp. 179–192). Cham: Springer.

Evans, G. (2003). Hard-branding the cultural city—from Prado to Prada. *International Journal of Urban and Regional Research*, *27*(2), 417–440.

Formato, R., & Presenza, A. (2018). *Management della destinazione turistica*. Rome: Franco Angeli.

García-Hernández, M., Calle-Vaquero, M. D. L., & Yubero, C. (2017). Cultural heritage and urban tourism: Historic city centres under pressure. *Sustainability*, *9*(8), 1346.

Go, F. M., Lemmetyinen, A., & Hakal, U. (2014). Introduction. In F. M. Go, A. Lemmetyinen, & U. Hakala (Eds.), *Harnessing place branding through cultural entrepreneurship*. London: Palgrave Macmillan.

Hall, C. M., Baird, T., James, M., & Ram, Y. (2016). Climate change and cultural heritage: Conservation and heritage tourism in the Anthropocene. *Journal of Heritage Tourism*, *11*(1), 10–24.

Jakob, D. (2012). The eventification of place: Urban development and experience consumption in Berlin and New York City. *European Urban and Regional Studies*, *20*(4), 447–459.

Kamel-Ahmed, E. (2015). What to conserve? Heritage, memory, and management of meanings. *Archnet-IJAR*, *9*(1), 67–76.

Lavers, T. (2012). 'Land grab' as development strategy? The political economy of agricultural investment in Ethiopia. *The Journal of Peasant Studies*, *39*(1), 105–132.

Liu, Z. (2003). Sustainable tourism development: A critique. *Journal of Sustainable Tourism*, *11*(6), 459–475.

Mackee, J., Askland, H. H., & Askew, L. (2014). Recovering cultural built heritage after natural disasters: A resilience perspective. [Article]. *International Journal of Disaster Resilience in the Built Environment*, *5*(2), 202–212.

McGuigan, J. (2005). Neo-liberalism, culture and policy. *International Journal of Cultural Policy*, *11*(3), 229–241.

Melè, D. (2003). The challenge of humanistic management. *Journal of Business Ethics*, *44*, 77–88.

Melè, D. (2009). Editorial introduction: Towards a more humanistic management. *Journal of Business Ethics*, *88*, 413–416.

Muresan, A., & Smith, K. (1998). Dracula's castle in Transylvania: Conflicting heritage marketing strategies. *International Journal of Heritage Studies*, *4*(2), 73–85.

Negussie, E. (2006). Implications of neo-liberalism for built heritage management: Institutional and ownership structures in Ireland and Sweden. *Urban Studies*, *43*(10), 1803–1824.

Nuryanti, W. (1996). Heritage and postmodern tourism. *Annals of Tourism Research*, *23*(2), 249–260.

O'Brian, D., & Miles, S. (2010). Cultural policy as rhetoric and reality: A comparative analysis of policy making in the peripheral north of England. *Cultural Trends*, *19*(1/2), 3–13.

O'Brien, G., O'Keefe, P., Jayawickrama, J., & Jigyasu, R. (2015). Developing a model for building resilience to climate risks for cultural heritage. *Journal of Cultural Heritage Management and Sustainable Development*, *5*(2), 99–114.
Picchio, M. (2019). Italian humanistic enterprise for social and environmental development: The vision of Brunello Cucinelli. *Studi Umbri*, 11(2). Retrieved from www.studiumbri.it/conoscenza/italian-humanistic-enterprise-for-social-and-environmental-development-the-vision-of-brunello-cucinelli/
Plaza, B. (2000). Evaluating the influence of a large cultural artifact in the attraction of tourism: The Guggenheim Museum Bilbao case. *Urban Affairs Review*, *36*(2), 264–274.
Ponzini, D. (2010). The process of privatisation of cultural heritage and the arts in Italy: Analysis and perspectives. *International Journal of Heritage Studies*, *16*(6), 508–521.
Power, D., & Scott, A. J. (Eds.). (2004). *Cultural industries and the production of culture*. London: Routledge.
Rocha, H., & Miles, R. (2009). A model of collaborative entrepreneurship for a more humanistic management. *Journal of Business Ethics*, *88*, 445–462.
Sakdiyakorn, M., & Sivarak, O. (2016). Innovation management in cultural heritage tourism: Experience from the Amphawa waterfront community, Thailand. *Asia Pacific Journal of Tourism Research*, *21*(2), 212–238.
Scott, A. J. (1997). The cultural economy of cities. *International Journal of Urban and Regional Research*, *21*(2), 323–339.
Soini, K., & Birkeland, I. (2014). Exploring scientific discourse on cultural sustainability. *Geoforum*, *51*, 213–223.
Spitzeck, H. (2011). An integrated model of humanistic management. *Journal of Business Ethics*, *99*, 51–62.
Starr, F. (2010). The business of heritage and the private sector. In S. Labadi & C. Long (Eds.), *Heritage and globalisation* (pp. 147–170). London: Routledge.
United Nations. (2013). *Managing cultural world heritage*. Paris: UNESCO/ICCROM/ICOMOS/IUCN.
Vecco, M. (2010). A definition of cultural heritage: From the tangible to the intangible. *Journal of Cultural Heritage*, *11*, 321–324.
Zetti, I. (2010). Built heritage, local communities and the production of territory. Citizen participation in heritage preservation. In M. Malkki & K. Schmidt-Thomè (Eds.), *Integrating aims—Built heritage in social and economic development* (pp. 231–250). Aalto: Aalto University.
Zoomers, A. (2010). Globalisation and the foreignisation of space: Seven processes driving the current global land grab. *The Journal of Peasant Studies*, *37*(2), 429–447.
Zukin, S. (1995). *The cultures of cities*. Malden: Blackwell Publishers.

7 Creative Tourism

A Humanistic Paradigm in Practice

Nancy Duxbury and Fiona Eva Bakas

1. Introduction

The growing humanistic paradigm provides a point of inspiration for transforming not only intra-organizational management practices but also shaping the macro dynamics of the intentions and operations of the tourism sector more widely. As Melé (2016) describes it, humanism

> sees the human being in permanent development and calls on him or her to flourish as a human. This is the responsibility of each human, but since the material, social and cultural environment can favor it, humanism seeks to foster the conditions for such flourishing including appropriate well-being development.
>
> (pp. 42–43)

Within the scope of tourism, the field of creative tourism appears well placed as a site of experimentation and a forerunner of humanistic practices and meaningful interactions between a locale's visitors and its residents. Creative tourism is characterized by four elements: active participation, a learning process, an opportunity for creative self-expression, and community engagement (Bakas, Duxbury & Castro, 2018). It is a type of experiential tourism, involving visitors as active participants, and evolves toward transformative travel, enabling creative self-expression and personal growth within the activities it develops and promotes. Creative tourism can be perceived as a step toward 'humanizing' travel as it includes an ethics of care for the locale in which activities are implemented as well as for the well-being and creative potential of the traveler.

Creative tourism emerged both as a development of cultural tourism and in opposition to the emergence of 'mass cultural tourism' (Duxbury & Richards, 2019). Going beyond tourism as a service industry and differing from conventional commercially driven tourism approaches, contemporary creative tourism is designed and implemented by local residents to promote interactions that foster collaborative creative expression and cross-cultural exchanges rather than pure economic transactions. For the communities in which it is developed, creative tourism can act as a driver for the revitalization of cultural traditions and forms, building on the embeddedness of the creative knowledge of artisan entrepreneurs (Richards, 2011) while re-vitalizing and sharing creative skills and engaging with the local community (Landry, 2010). In rural and small-city contexts, culture-based revitalization through creative tourism can stimulate and build community cohesion and contribute to holistic and sustainable community development (Duxbury, Campbell & Keurvorst, 2011).

This chapter examines and interprets leading practices in contemporary creative tourism in the light of a humanistic approach to tourism, with a particular focus on the rural and small-city development context. It is informed by accumulated knowledge, observations, and reflections on the development of creative-tourism activities within a 44-month (2016–2020) national research-and-application project, CREATOUR, which has been catalyzing a network of 40 creative-tourism initiatives (referred to here as 'pilots') located in small cities and rural areas throughout Portugal. Focusing on the values, strategies, and actions of eight pilot organizations within this project, this chapter considers the following: How do creative-tourism strategies and practices embody, operationalize, and advance a humanistic paradigm? How does a humanistic management perspective produce new ideas about creative tourism relative to more standard approaches?

This chapter begins with an overview of four key dimensions of humanistic management approaches that resonate strongly with the objectives and goals of creative tourism: (1) promoting human flourishing, (2) engaging the other in a journey of mutual discovery, (3) honoring the dignity of each stakeholder through inclusion in decision-making, and (4) contributing to the common good. It presents an overview of creative tourism and the CREATOUR project and then outlines the methodology used. The core of the chapter examines and illustrates how these four dimensions are adopted in the values, strategies, and practices of creative-tourism organizations. In closing, the chapter reflects on the implications of using a humanistic management paradigm as a new perspective on understanding creative tourism.

2. Humanistic Management

While there is "not unanimous consent on what humanistic management means" (Melé, 2016, p. 39), Dierksmeier's (2016) definition is a useful touchstone: "an unconditional commitment to orient business at the intrinsic worth of human life, that is, a reconceptualization of business as being at the service of human dignity and the flourishing of human life" (p. 27). It puts humans central to development and the economy—the whole human, including their learning and creativity and seeking for self-actualization and meaning. The principles and values discussed within humanistic management can be adopted to examine, address, and re-configure local dynamics and interactions with regard to tourists.

2.1 Promote Human Flourishing

Human flourishing is related to the sustainable pursuit of self-actualization and fulfilment within the context of a larger community of individuals, each with the right to pursue his or her own such efforts (Little, Salmela-Aro & Phillips, 2014). It encompasses the uniqueness, dignity, diversity, freedom, happiness, and holistic well-being of the individual within the larger family, community, and population (Ryff & Singer, 2008). This aligns well with the eudaimonic meaning of well-being, "the realization of someone's potential" (Kabadayi, Alkire, Broad, Livne-Tarandach, Wasieleski & Marie Puente, 2019), which captures the essence of the two great ancient Greek imperatives: first, to know yourself, and second, to become what you are (Ryff, 2014).

Creativity and artistic activities have repeatedly been shown to be important to human development and flourishing, that is, 'to becoming fully human' (Wright & Pascoe, 2015), enhancing self-understanding, self-fulfillment, and self-actualization (Berman, 1998; Langer, 2005) as well as health and well-being (Clift & Camic, 2016). The arts also have a central role in teaching critical thinking, including the ability to acquire capacities for human empathy (Nussbaum, 2010).

Opportunities for creative inspiration and personal reflection—"specific practices that allow for aspects of 'self' to be expressed, reflected upon, and cared for in different ways" (Steckler & Waddock, 2018, p. 191)—are important for sustaining oneself over time. Yet "having the opportunity to exercise your creativity . . . to acknowledge your life experience . . . these are things that are not necessarily a part of people's everyday lives" (artist jil p. weaving, in Kallis, 2014, p. 241). Steckler and Waddock (2018) found that retreats provided social entrepreneurs essential space "to clear their thoughts, facilitate other ways of viewing the world and themselves, and find new sources of inspiration" (p. 192). Nearly half of the respondents in their study engaged in "various forms of inspirational retreats, which were often associated with the arts or other aesthetic ('beauty') appreciation" (p. 191). Links among creative self-expression, well-being, and a desire for personal transformation inform the evolution of creative tourism today (Duxbury, Kastenholz & Cunha, 2019).

2.2 Engage the Other in a Process of Mutual Discovery and Dialogue to Learn About Others and About One's Self

Alongside *well-being*, the idea of *being well* is defined as "the empathetic, successful and gratifying relationship of a person with others, with nature and with the whole" (Natura, 2009). As Pirson (2018) observes, we need to

> reclaim our humanity and dignity through a novel civility that . . . respects our intrinsic value as human beings and allows us to engage freely with each other based on love and compassion. . . . A civility that respect[s] the fact that we are unconditionally worthy and . . . life forms that wish to flourish and thrive.
>
> (p. 104)

Discussions about human dignity within humanistic management stress the importance of a meaningful collaboration that requires the parties to develop trust through dialogue (Rodríguez-Lluesma, Davila & Elvira, 2014). In a collaborative context, the *third room* refers to a metaphorical shared space in which

> all participants bring their experiences, ideas and skills to share, creating something that could not happen without the energy and ability of everyone present. . . . the third room defines the unexpected opportunities that collectively we can discover when we join our efforts.
>
> (Kallis, 2014, p. 59)

Creative activity can "function both as a meditative or contemplative personal time but also as a framework for building a community-connecting third room" (Kallis, 2014, p. 113). The agency of creative action for personal growth tends to foster an openness to connecting with others. As community-engaged artists know, "when people's hands are busy their minds and mouths will open" (Kallis, 2014, p. 80, citing artist Marina Szijarto). Artist jil p. weaving has experienced this in many collaborative community-based projects and observes:

> When you will feel like you are fully participating in your own life, exploring your creativity and sharing your life experiences, that is when we are most human and that is when we connect with each other. So providing the opportunity for people to be their best selves can create a world where people want to connect with each other.
>
> (cited in Kallis, 2014, p. 241)

In the travel context, such connecting is centered on cross-cultural exchange and co-learning. Providing a creative platform or *third room* to foster these connections can move simple exchanges toward processes of co-creation and collaborative meaning-making in which participants engage in a mutual discovery journey.

2.3 *Honor the Dignity of Each Stakeholder by Involving Them as Much as Possible in Decisions That Impact Their Lives*

Humanistic businesses aim to make products that address genuine human needs and do it in ways that respect the concerns of all stakeholders. Beyond this, humanistic management aims to involve all stakeholders in decision-making as much as possible in order to honor the people who are affected by any changes that decisions and courses of action (e.g. tourism development) may have on their lives (Dierksmeier, 2016). For example, creative tourism mediator-entrepreneurs who link artisans to tourism in Portuguese small cities and rural area contexts embody a deep sense of responsibility toward the artisans they work with, which is embedded in their conceptualization of entrepreneurship and their practices of involving local artisans and a broader network of community actors in planning the activities (Bakas, Duxbury & Castro, 2018). These micro-practices align with the humanistic notion of society as a group of free people ruled by justice and benevolence, living together, acting with reciprocity and cooperation, without losing their personality, thus maintaining their cultural and individual distinctiveness (Melé, 2016).

2.4 *Share the Common Goods of Society and Contribute to the Common Good in Some Way*

The field of social entrepreneurship holds the promise to identify humanistic business models that are driven by the principle of contributing to the common good: "Social change agents, or social entrepreneurs, are passionate leaders who generate and drive ideas and who aspire to make a positive and impactful difference" (Steckler & Waddock, 2018, p. 172). Social entrepreneurs prioritize human and societal well-being and the advancement of humane and life-conducive organizing. They are mindful of accomplishing social and economic objectives in a balanced way (Chinchilla & Garcia, 2017), and to contribute to the wider society and place in which they are embedded.

One of the socio-cultural contributions that artists and creative-tourism entrepreneurs make is the recovery and relearning of almost-lost skills, "*un-venting* ancestral knowledge and technologies we are otherwise at risk of losing" (Kallis, 2014, p. 21, emphasis added). Many creative-tourism initiatives in rural areas are focused on the revitalization of traditional craft skills and aesthetics of place (Duxbury, Silva & Castro, 2019). Contributing to the common good of a community can also involve passing on cultural knowledge and artistic skills. For example, rural tourism handicraft entrepreneurs in rural Greece offer, for free, services to rural communities that are otherwise missing, such as felting lessons to children (Bakas, 2014).

3. Creative Tourism and the CREATOUR Project

Creative tourism provides a nurturing milieu for linking cultural and creative practices with a humanistic approach to tourism. In rural and small-city contexts, it promises to innovate alternative tourism pathways, small scale by nature but incrementally inspirational and rhizomic in its potential. Creative tourism creates a closer, egalitarian relationship between tourists and residents, which derives from the immersion of both in local culture through active participation in creative learning experiences. It privileges humane and respectful interactions and seeks to foster new dynamics among hosts, local residents, and visitors. The design and development of creative-tourism offerings within this framework places great attention on local meaningfulness and culture- and place-based inspiration. It explicitly engages with important issues of cultural vitality, sustainability, and exchange through creative practices that are enrooted in specificities of place, culture, and ways of life.

The demand for creative tourism is driven by travelers seeking more active and participative cultural experiences in which they can use and develop their own creativity. There is an increasing realization that acquiring material objects or 'random' experiences is a form of empty consumerism and that happiness is better pursued though transformative experiences that focus on mindful self-development (Wolf, Ainsworth & Crowley, 2017). This phenomenon is giving rise to transformational tourists who are, in effect, looking for existential authenticity, a special state of being in which one is true to oneself. Ross (2010) defines transformative travel as "sustainable travel embarked upon by the traveler for the primary and intentional purpose of creating conditions conducive for one or more fundamental structures of the self to transform" (p. 55). Self-expression through creative activities is seen as a way of articulating one's authentic self and getting closer to this transformation (UNWTO, 2016). The tourist in this context is an active, meaning-seeking, creative traveler—a fully thinking and striving human, not just an economic agent/customer. The hosts and residents of the visited community are similarly 'real people' with diverse personalities, life narratives, knowledges, skills, opinions, challenges, and aspirations.

CREATOUR is a 44-month (2016–2020) interdisciplinary research-and-application project that is developing a network of creative-tourism initiatives for the first time in Portugal.[1] It involves five research centers working with 40 participant organizations (pilots) offering creative-tourism initiatives in small cities and rural areas across Portugal in the Norte, Centro, Alentejo, and Algarve regions (Figure 7.1). The CREATOUR project's design, strategy, and implementation axes focus on human interaction, exchange, dialogue, and creation as platforms for learning and actualization. CREATOUR provides its participants with

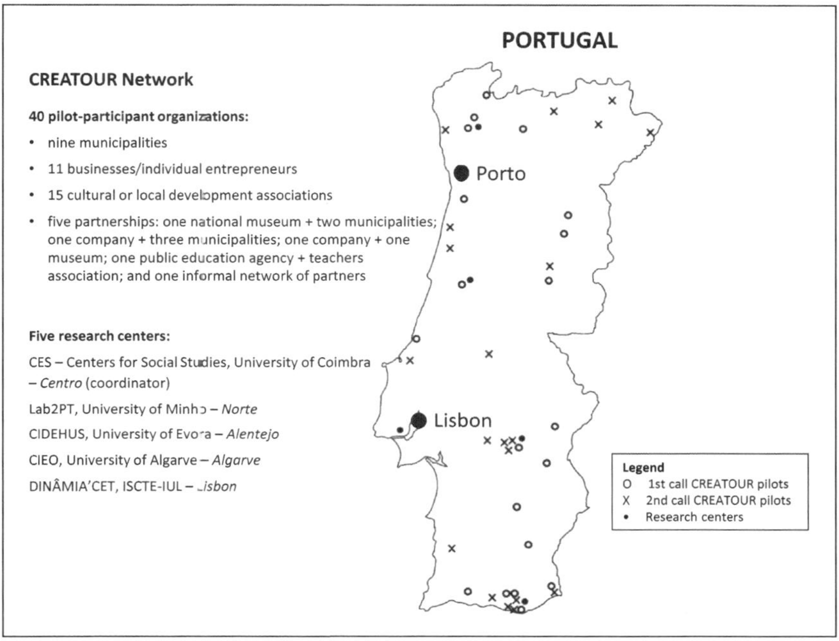

Figure 7.1 Types of CREATOUR pilot organizations (Portugal)

Source: Authors' own elaboration

opportunities to build and share knowledge, network with others, design and improve creative-tourism offers, and create strategies to enhance community benefit. Within a research-and-application context, the pilots strive to develop sustainable creative-tourism initiatives and enhance the socio-cultural and economic dimensions of their local communities.

Within the CREATOUR project, creative-tourism activities based on local traditions, artistic expression, and place began appearing in the four Portuguese regions from June 2017. These creative-tourism activities include:

- Craft workshops—textile, pottery, ceramics, leather, metal, wood, etc.
- Fine arts workshops—painting, sculpture, drawing, and illustration
- Photography, video, and digital arts workshops
- Performing arts workshops and community-engaged, participatory artistic residencies
- Storytelling sessions and workshops
- Gastronomy-focused workshops

- Creative and interpretative 'Walks & Visits' involving creation activities
- Ancestral traditions workshops and active-participation activities
- Raw materials production and work cycles—of salt, linen, wool, clay, marble, wicker, etc.

Each project was imagined, designed, and implemented locally, with a diversity of approaches, cultural traditions, and place specificities embedded within each initiative.

4. Methodology

Qualitative knowledge co-created with eight CREATOUR creative-tourism managers is used in this chapter to analyze the ways in which creative-tourism managers' strategic approaches to creative tourism can be viewed as adhering to a humanistic paradigm. Within the project, 24 semi-structured interviews were held with creative-tourism managers at annual on-site visits during 2017, 2018, and 2019. Participant observations, informal conversations with creative tourists, and lived memories from on-site visits as well as a number of discussions and informal conversations with creative-tourism managers, enabled through their participation in a series of IdeaLabs within the project, also inform this chapter. These interactive, two-day focus group/workshop/mentoring sessions involved creative-tourism managers and researchers, and occurred three times annually. The two researcher-authors of this chapter conducted a reflexive process (Cheek, Lipschitz, Abrams, Vago & Nakamura, 2015) on their ethnographic experience immersed within this three-year research-and-application project to select and analyze data for this chapter. This reflexive process informed the choice of subject and discussion of data in an overarching way.

The main characteristics of the eight CREATOUR creative-tourism pilot organizations selected for analysis are presented in Table 7.1 to provide some background information about these entities. The ways in which the views and strategic decisions of creative-tourism managers can be perceived as humanistic are thematically analyzed using the humanistic framework described in the introduction, interweaving their statements over a three-year period with related literature.

5. Findings and Analysis

Using the four dimensions of humanistic management presented earlier as a framework, and focusing on eight CREATOUR pilot organizations, this section examines how their creative-tourism strategies and practices embody, operationalize, and advance a humanistic paradigm. Insights from the managers of these organizations are discussed, with a summary of key points in this chapter presented in Table 7.2.

Table 7.1 Description of CREATOUR pilot organizations discussed in this chapter

CREATOUR pilot	*Location and region*	*Description*
Cerdeira Village (ADXTUR network)	near Lousa (Centro)	Luxury rural rental accommodations and artistic studio spaces in a deserted mountain village, managed by a German sculptor who helped restore the village over the past 30 years. Artistic residencies and artisan workshops for tourists on a variety of arts, including woodcarving and pottery, are provided in cooperation with the tourism development agency of the Schist Village Network, ADXTUR.
Encontrarte Amares	Amares (Norte)	This small-scale, five-day, biennial arts festival aims to connect contemporary arts with local tradition through promoting artistic experimentation. The festival also promotes the regions' immaterial heritage, highlighting local gastronomy, customs, images, sounds, stories, and know-how.
Estival	near Guarda (Centro)	An annual, weeklong art festival held at a farm owned by a Dutch couple in the mountainous interior. About 80 participants engage in workshops such as theater, stand-up comedy, singing, dancing, and wood sculpture. On one evening, participants travel to the closest village to eat and dance with local residents, and festival artists perform for them.
L Burro i L Gueiteiro	Miranda do Douro (Notre)	An annual itinerant festival that travels between three villages. This alternative traditional festival serves as a way of preserving the breed of traditional donkeys of Miranda as well as local culture. Workshops offered during the festival focus on traditional music and dances of the region and on playing traditional instruments, in particular, the bagpipe.
Proactivetur	Loulé (Algarve)	As a tour operator, Proactivetur offers half-day or full-day workshops on ancestral craft techniques such as cane basket weaving, palm weaving, and making traditional floor tiles. It is also a project manager of the TASA project (Ancestral Techniques Current Solutions) which aims to bring strategic innovation to the craft industry by encouraging ancestral craft techniques to be used in modern product design.
Quico Turismo	Nazaré (Centro)	Quico Turismo is primarily a room-rental business that hosts visitors, expanding to also offer photography expeditions with a cultural-historic focus and workshops sewing keyrings in the shape of *carapau* (horse-mackerel) fish, which are traditionally associated with Nazaré. These activities form the initial steps in the development of Nazaré Criativo.
VERde NOVO	Ribeira da Pena (Norte)	A tourism and culture development consultancy, VERde NOVO offers creative-tourism workshops and guided tours on the cycle of linen—from sowing to weaving—in line with the traditional practices of this village.
VIC Aveiro Arts House	Aveiro (Centro)	VIC combines a local-themed tourist accommodation space, an artistic residence and co-work space, a cultural space with a small auditorium, and a gallery where cultural events and training events take place. Creative-tourism activities have included workshops in video production and editing, construction of sonorous sculptures and of 'noise puppets', and silk-screening.

Source: Authors' own elaboration

Table 7.2 Enacting humanistic management principles in creative tourism

Humanistic management principle	*Expression of humanistic management in action*	*Related creative-tourism strategies and actions*
Promote human flourishing	Promote well-being	Design activities that aim to boost tourists' well-being through mindful and transformative experiences.
		Facilitate visitor–local interactions to increase locals' well-being in a manner that is not hyper-commercialized.
	Foster self-awareness toward self-actualization	Provide opportunities for creative self-expression.
		Encourage and enable greater self-knowledge for tourists and local residents.
	Foster social connectedness	Design activities to provide opportunities for creating good feelings and connectedness, which contributes to building social capital, especially in the context of small art festivals.
Engage the other in a process of mutual discovery and dialogue	Intergenerational mixing and exchange	Mix all ages of people during creative-tourism workshops, creating space for meaningful exchange and dialogue while making, co-learning, and helping one another respectfully.
	Think in new ways through cross-cultural dialogue	Encourage visiting artists involved in creative-tourism workshops to be influenced by local residents, tourists, and other artists they meet while on location.
Involve stakeholders in decisions that impact their lives	Expand and directly engage the range of stakeholders considered	Imagine and invent approaches to diversely involve a wide range of locals and tourists, including elderly people and children, who are often forgotten within tourism.
	Active participation of artisans in planning tourism product	Involve local artisans in planning creative-tourism products from the beginning.
	Protection of dignity of human life	Consult and involve elderly residents as stakeholders in ways that promote and protect their dignity.
Contribute to the common good	Add value to the society at large	Intentionally foster imaginative new ways to connect place and communities and revitalize rural traditions and intangible culture.
	Act with reciprocity	Give back to the community where the creative-tourism activities take place through free workshops, cultural shows, and other complementary activities.
	Managers deal with social responsibility and profitability at the same time	Consciously adopt social economy principles, particularly in terms of rationales to operate 'for the good' of the local community.
		Continuously aim to balance social responsibility and profitability.

Source: Authors' own elaboration

5.1 Promoting Human Flourishing

Promoting human flourishing is a humanistic management goal that can be achieved by increasing well-being, fostering self-awareness, and increasing social connectedness. *Increasing well-being* features prominently in creative-tourism managers' aspired outcomes from their work. The manager of Cerdeira Village believes that the process of making things generates well-being and provides "powerful satisfaction" for those making (interview, 13/03/2018). According to Aristotle, we become via doing, meaning that the very act of creating something brings us closer to positive development of self. An increase in well-being also occurs when individuals are involved in collective creative activities. In the case of a documentary-making project developed within the Encontrarte Amares festival, participating residents' well-being was visible: "It's fantastic to see the video in which the old men in the city are singing and dancing. This is extraordinary" (interview, manager of Encontrarte Amares, 01/08/2017). Fostering good feelings is a very humanistic goal to achieve.

Creative-tourism managers speak about how *fostering self-awareness* in local residents is important to them. Creative tourism can contribute to building and highlighting the value of people and their work; as a manager of VERde NOVO says, "Creative tourism helps value people, products, and territory" (interview, 15/03/2019). Creative tourism presents one's home-place to travelers in a meaningful manner, respectful of the existing traditions, while new artistic performances/interventions can inform how a landscape is encountered and represented and can create a renewed sense of belonging. A narrative of self can emerge through an emotional interaction with the landscape (Miller, 2017). Part of the Estival organizer's philosophy is that creative activities encourage people to see things from a new perspective and create a space for local residents to acknowledge and reflect upon their life experience.

Fostering *social connectedness* is another humanistic management element that creative-tourism managers strive to foster within their activities. As a manager of VERde NOVO says, "Creative tourism is also about people cooperating and interacting and people growing as citizens" (interview, 15/03/2019). VIC Aveiro Arts House focuses on interconnecting the creative people living in its small city. When VIC's manager spoke to creative people in Aveiro, he found that "people engaged in creative activities felt isolated" and wanted to help them. He made it one of his key objectives to connect these local creators and encourage them to interrelate: "Our main objective was to get these people together, to be together and to start naturally talking and see what comes out of this" (interview, 17/04/2019). VIC organizes international artistic residencies and creative workshops that provide platforms for connecting and interacting around shared interests and ideas. Similarly, the manager of Quico Turismo encourages the artisans in her micro-network to interact with one another by inviting them to each other's workshops; she feels that this interaction and bonding will "help create a community of creative people" (interview, 13/03/2018).

Small-scale festivals put a significant emphasis on developing a milieu that encourages social connectedness among visitor-participants and local residents. An important component of the L Burro i L Gueiteiro festival is its social connectedness and friendly atmosphere. As the festival organizer says, "Participants even offer their help voluntarily to complete the daily tasks of the festival and it is this interaction, in fact, that later translates into a very friendly and 'familiar' ambiance" (site-visit interview, 2018). Within Encontrarte Amares, social connectedness is generated through participants, organizers, artists, and foreign tourists all working together. As the manager says, "people were there and did not even realize who was the organization, the tourist and the participant. That is, everyone did everything and collaborated. They had a participatory and collaborative spirit" (interview, 01/08/2017). The creation of environments to foster these collaborative practices is a strategic position of the managers (Bakas, Duxbury, Remoaldo & Matos, 2019) and adheres closely to the humanistic notion of society as a group of free people ruled by justice and benevolence, living together, acting with reciprocity and cooperation (Melé, 2016).

5.2 Engage the Other in a Process of Mutual Discovery and Dialogue to Learn About Others and About One's Self

Humanistic management stresses the importance of meaningful collaboration that requires the parties to develop trust through dialogue. Engaging in creative-tourism activities provides a safe and friendly space for this to occur. In the act of collective doing, dialogue can easily be stimulated. Creative-tourism activities effectively create a 'third room' where all participants bring their ideas to the table and share them, "creating something that could not happen without the energy and ability of everyone present" (Kallis, 2014, p. 59).

Creative-tourism workshops seem particularly good at catalyzing a mixing of people from various age groups and thus can promote *intergenerational mixing* and exchange in ways that foster dialogue and mutual discovery to learn about oneself and others. As the Estival organizer says, "Young and old people mix really well in the workshops. . . . They are helping each other in the workshops. This is very interesting as they try to create a community there. Everyone is valued and respecting" (interview, 09/08/2018). Similarly, the organizers of the L Burro i L Gueiteiro festival stress the significance of interaction among various age groups:

> The interaction amongst the participants is something that distinguishes this festival from others because it is a very friendly festival. It is friendly because we are able to involve both kids and adults in

> our activities, that is, we work with age groups from the baby of months to people of 70, 80, 90 years.
>
> (interview, 29/07/2018)

Encounters with 'others', which may be a person of a different age group, constitute a reminder to engage the other in a process of mutual discovery and respectful dialogue.

Creative tourism also promotes cross-cultural dialogues that can stimulate *new ways of thinking* in both local residents and visiting artists and tourists. The manager of VIC Aveiro Arts House has observed that visiting artists who present a workshop are influenced by this interactive work while, at the same time, the discussions they hold and the work they conduct in Aveiro can affect the way the place is (re)conceived by locals. Speaking about a visiting writer, he says:

> The artist affects the city, the city affects the artist, the creation. The artists bring something that is in his head but it will change depending on his experience in the place, the dynamic of the city and the experience he lives there. He wrote two texts about Aveiro that were very insightful.
>
> (interview, 17/04/2019)

For local residents participating in creative-tourism events, contact with artists and other participants from other countries—or just other parts of the country or walks of life—can stimulate new ways of thinking. Within Encontrarte Amares, the manager observes this happening: "There is information that is being given, new ways of thinking, new ways of acting, they [local participants] begin to problematize their day-to-day and to think with new ways of dialoguing and thinking about their reality" (interview, 01/08/2017). Thinking differently also results from local communities' contact with visitors in creative-tourism workshops and related events during the L Burro i L Gueiteiro festival. The organizers believe these activities are creating additional cultural experiences and meanings for the villages where the festival takes place, providing opportunities for local communities to develop a sense of shared experiences, values, and belonging, and helping to counter exclusion in these rural and remote areas.

5.3 Protect the Dignity of Each Stakeholder by Involving Them as Much as Possible in Decisions That Impact Their Lives

Creative-tourism entrepreneurs within the CREATOUR project place great importance on engaging a variety of stakeholders, especially the local community, often involving them in planning processes as well as final activities. At Encontrarte Amares, the *community is involved* throughout the festival program, with an accent on fostering "community empowerment" and active participation of local residents throughout the program. The event's design explicitly promotes meaningful connections between the local community and the artists performing in the festival; as the festival manager says, "We have always had a dialogue, in all projects, between the local community and the artists" (interview, 01/08/2017).

Directly involving the local community within events that take place in festivals provides a platform of a higher value than to be involved only as 'helpers' or the people that festival-goers interact with on a commercial basis. Along these lines, Encontrarte Amares *promotes the dignity* of all stakeholders by directly involving cafeteria workers in a performance at the festival. The festival manager describes this process:

> Ekatrin is a Russian artist, who works with art and community. We set a challenge for her to work and follow the whole process of the cafeteria. This was a participation and a co-creation exercise, where the ladies shared their daily life . . . they also created a performance on the last day of the festival . . . about who they are and what they do.
>
> (interview, 01/08/2017)

In the upcoming festival, the plan is to involve the cafeteria ladies in a ballet. All people are creative, but they need a fostering milieu to discover and practice their creativity. These creative interactions allow the cafeteria workers to accomplish this while fostering self-esteem and non-hierarchical societal relationships, thus achieving a greater degree of humanism in the festival.

The manager of Cerdeira Village also speaks of the importance of a relationship to the local community, a resource that embodies and expresses culture: "Community *is* culture" (interview, 13/03/2018). Many of the creative-tourism activities being developed within CREATOUR tap into this wellspring of local cultural traditions, skills, and aesthetics of place. Consequently, *local artisans are stakeholders* involved in the planning of creative-tourism workshops, which often include elderly members of the community who possess traditional knowledges and skills. The manager of Proactivetur says: "The artisans are a fundamental part of the local culture and authenticity we want to bring in the programs. When we know them and what they do, we can discuss how to share some of that knowledge in a creative experience" (interview, 15/03/2018). The manager of Quico Turismo takes a collaborative approach to planning creative-tourism workshops: she presents an idea to the artisans and then they discuss it, making alterations and suggestions of how to materialize the idea. For example, they did a test run of the *carapau* sewing workshop to see what they would need for the activity—this was a joint idea of Celia and elder artisan Fatima (interview, 13/03/2018). VERde NOVO usually takes the initiative to define the creative-tourism product concept and sketch out how the activity should run and then presents it to the artisans, asking them for feedback, whether they want to participate in the activity, and how they would like to do it.

Creative-tourism managers also try to include *children* in their activities. For example, the proprietor of Quico Turismo has created a book for children on the local history of Nazaré, and the proprietor of Cerdeira Village has planned artisan workshops and an art exhibit for local fourth-grade students to coincide with International Children's Day. As the proprietor of Cerdeira Village says, "Children are part of the culture that is forgotten" (interview, 05/02/2018). Artisans in rural areas often provide early training grounds for local young people in the absence of the kinds of infrastructure available in the city (Luckman, 2012; Bakas, 2014). Involving the often-neglected stakeholders of tourism (Bakas, 2018) in planning their activities, creative-tourism managers attempt to protect the dignity of all stakeholders and give them a voice—a humanistic approach to tourism management.

5.4 Share the Common Goods of Society and Contribute to the Common Good in Some Way

Many creative-tourism managers adhere to the core belief that the *local society should benefit* from their economic activity in some way. The design of the CREATOUR project also embodies this ambition. Many creative-tourism managers within the CREATOUR project enact principles of social entrepreneurship by prioritizing human and societal well-being and contribute to the wider society and place in which they are embedded (Chinchilla & Garcia, 2017). For example, the VERde NOVO managers stress how their main goal is "to build value, not just financial value—activities should also build cultural and social value" (interview, 15/03/2019).

Acting with reciprocity by giving back to the community is integral to many of the creative-tourism initiatives. For example, as part of her creative-tourism entrepreneurial actions, the proprietor of Quico Turismo made a documentary with the municipality on the fishermen in Nazaré because she is "interested in promoting the culture of the area and helping the local community" (interview, 13/03/2018). Bringing cultural activities into a community is the flipside of this coin. For example, the L Burro i L Gueiteiro festival benefits local communities by offering a type of entertainment that is lacking in the area. The festival manager says "the feedback from the village communities is extremely positive because these communities are quite devoid of this type of cultural event". By developing new cultural opportunities and socio-cultural experiences, such creative-tourism initiatives are contributing to community well-being and enhancing cultural vibrancy.

The managers of VERde NOVO hold a *deep sense of responsibility* toward the artisans they work with in terms of helping them sell their items and improving their economic situation—this is something that is embedded within their conceptualization of entrepreneurship. As they say, "It's not in our DNA to just do tourism animation and leave! We want to be part of the community and get public institutions to work with us. We want to run a community entrepreneurship program" (interview, 12/03/2018). The manager of VIC Aveiro Arts House stresses how important it is that the artists get a fair wage for their efforts and sometimes dips into his own profits to pay artists if the creative-tourism workshop has not generated enough funds, illustrating a very humanistic concern for the well-being of others. As he says:

> If the people are not enough for the workshop, and if the person who is holding the workshop is not getting enough from the inscriptions, we put in a bit more money so they can have a fair pay for what they did.
>
> (interview, 17/04/2019)

In these ways, many creative-tourism managers are mindful of accomplishing social and economic objectives in a balanced way. Managers of creative-tourism workshops often mention how they deal with *social responsibility and profitability at the same time*. Their conceptualization of their creative-tourism activity often goes beyond purely profit-making terms, and they include social responsibility in their discourse. For example, the manager of VIC Aveiro Arts House is creating a series of creative-tourism workshops that could be a meeting point for locals, the local temporary population (e.g. students and researchers), and tourists to exchange knowledge, ideas, and experiences. While conscious of the need to keep the overall enterprise operational, he is more concerned about creating deep relationships among the local population: "The ideal would be to transcend commercial dynamics—the relationships that you develop within a non-economic context are much deeper and more interesting" (interview, 17/04/2019). Looking forward, he plans to create cultural management and curatorship workshops, not only to draw people from other parts of the country, and thus make a profit, but also to involve locals and empower them to lead events programming in the community (interview, 17/04/2019).

6. Closing Reflections

The niche area of creative tourism is becoming increasingly significant, especially in extra-metropolitan areas, for fostering significant 'soft' impacts such as increased community engagement and cultural sustainability (Duxbury & Richards, 2019). Creative tourism creates platforms for meeting, exchange, co-creation, and learning—bringing together traveling humans and locally-resident humans to share skills, stories, and creative processes for mutual enrichment and benefit. Creative tourism is a type of tourism that can meaningfully contribute to strategies for "full-spectrum flourishing" (Laszlo, 2019, p. 1), defined as a world in which people and all life thrive now and across future generations, which can be strengthened by humanistic approaches.

Currently, the creative-tourism field is characterized by a wide array of smaller-scale activities and initiatives designed and offered by small entrepreneurial enterprises, non-profit cultural and local development associations, local and regional government agencies, and individual artists and creators. This rich diversity of organizations and activities is a wellspring for considering, implementing, and evolving new approaches in tourism, including the development of strategies and practices that are inspired by and follow principles of humanistic management and look much further than the economic bottom line of their operations. Practitioners, however, are often separated and pursuing initiatives in parallel, and linking these efforts together to inform both research and practice forms an essential step in identifying commonalities, variations, and possible trajectories forward.

Focusing on eight creative-tourism pilot organizations within the CREATOUR project, this chapter examined their creative-tourism management strategies and practices from a humanistic perspective. Four key dimensions of humanistic management were analyzed: (1) promoting human flourishing, (2) engaging the other in a journey of mutual discovery, (3) honoring the dignity of each stakeholder through inclusion in decision-making, and (4) contributing to the common good. We used these dimensions as a framework to examine and illustrate how they are adopted in practice in creative tourism, with a particular focus on the rural and small-city context. A humanistic lens provided an insightful light that illuminated and interlinked strategic features and intentions we have observed in practice.

While principles of humanistic management can be found in multiple fields, we believe the compilation of approaches found in the eight organizations examined here collectively articulate a humanistic approach within creative tourism. The management strategies of the creative-tourism entrepreneurs presented in this chapter blend social entrepreneurship values and approaches with concern for the cultural vitality and sustainability of their local community and the well-being of both local and visitor participants in their activities. The activities they design intentionally provide opportunities for learning, exchange, and creative self-expression. The creative processes at the heart of these activities form the basis for participants to gain greater self-knowledge, boost well-being, and foster mindful and potentially transformative experiences.

Strategies to encourage mixing, interaction, and thoughtful engagement are tightly linked with the design of these creative activities. The entrepreneurs aim to facilitate meaningful visitor–local interactions through designing activities that provide opportunities for creating good feelings, connectedness, and building social capital. The mixing of diverse participants of different ages during creative-tourism workshops creates a space for meaningful exchange and dialogue that occurs while making and helping one another, and encourages openness to mutual co-learning and co-influence.

The initiatives developed are integrally immersed in the local context, culturally and socially. Local artisans are usually involved in planning creative-tourism products from the beginning. Elderly residents are consulted and involved as stakeholders in ways that promote and protect their dignity. Approaches are invented to diversely involve a wide range of locals and tourists, including elderly people and children, who are often forgotten within tourism.

These entrepreneurs follow social economy principles, particularly in terms of rationales to operate 'for the good' of the local community and to continuously aim to balance social responsibility and profitability. As culture-based initiatives, creative tourism also provides an avenue to 'give back' to the local cultural 'richness' that inspires and informs these activities. In the short term, they give back to the communities where the creative-tourism activities take place through free workshops, cultural shows, and other complementary activities. In the medium and longer term, they intentionally foster imaginative new ways to connect place and communities and revitalize rural traditions and intangible culture.

This chapter provided a new lens to advance thinking about creative tourism, which is usually presented and contextualized as a meeting place at the intersection of 'creativity' and 'tourism' (Richards, 2011). Little attention has been paid to the operational strategies involved in developing and implementing creative-tourism activities, or to the broader intentions of creative-tourism entrepreneurs in pursuing this work. This chapter highlighted how creative-tourism practices can embody a humanistic paradigm. A humanistic perspective on creative tourism helped reveal and articulate the underlying values, implementation strategies, and greater purposes to which creative tourism can contribute. We modestly hope that this chapter contributes to demonstrating how creative tourism can be a platform for advancing social entrepreneurship and humanistic practices in tourism.

Creative tourism can contribute to human-centered development. As creative tourism evolves to become ever more aligned with individual and collective desires to undertake travel that is personally transformative and socially minded, creative experiences can be strategically designed as either retreat spaces—to take time out to refresh or to find inspiration, greater self-awareness, and potential transformation—or as platforms for furthering collective action within a humanistic context. The power of culture and the arts to inspire, inform, and act as platforms for renewed values, competencies, and behaviors for more holistically sustainable living is increasingly recognized in research (Dessein, Soini, Fairclough & Horlings, 2015; Kagan, 2011), policy (Duxbury, Hosagrahar & Pascual, 2016), and practice (Riccardi & Ferreira, 2019). In this context, possible alliances among active creative activity and expression, tourism and travel, and advancing the principles and practices of humanistic management deserve further research and experimentation in practice.

Acknowledgements

This chapter was developed within the framework of the research project 'CREATOUR: Creative Tourism Destination Development in Small Cities and Rural Areas' (Project 016437), which is supported by the Portuguese Foundation for Science and Technology (FCT/MEC) through national funds and co-funded by FEDER through the Joint Activities Programme of COMPETE 2020 and the Regional Operational Programmes of Lisbon and Algarve. The authors thank the interviewed participants for their precious time and valuable insights and the CREATOUR team at Lab2PT for conducting background interviews with the managers of the L Burro i L Gueiteiro and Encontrarte Amares festivals.

Note

1. For more information about the CREATOUR project, see www.creatour.pt.

References

Bakas, F. E. (2014). *Tourism, female entrepreneurship and gender: Crafting new economic realities in rural Greece* (PhD thesis), Otago University, New Zealand. Retrieved from http://hdl.handle.net/10523/5381.

Bakas, F. E. (2018). The political economy of tourism: Children's neglected role. *Tourism Analysis*, *23*(2), 215–225. doi:10.3727/108354218X15210313504562

Bakas, F. E., Duxbury, N., & Castro, T. V. de. (2018). Creative tourism: Catalysing artisan entrepreneur networks in rural Portugal. *International Journal of Entrepreneurial Behavior & Research*, *25*(4), 731–752. doi:10.1108/IJEBR-03-2018-0177

Bakas, F. E., Duxbury, N., Remoaldo, P. C., & Matos, O. (2019). The social utility of small-scale art festivals with creative tourism in Portugal. *International Journal of Event and Festival Management*, *10*(3), 248–266. doi:10.1108/IJEFM-02-2019-0009

Berman, H. J. (1998). Creativity and aging: Personal journals and the creation of self. *Journal of Aging and Identity*, *3*, 3–9.

Cheek, J., Lipschitz, D. L., Abrams, E. M., Vago, D. R., & Nakamura, Y. (2015). Dynamic reflexivity in action: An armchair walkthrough of a qualitatively driven mixed-method and multiple methods study of mindfulness training in schoolchildren. *Qualitative Health Research*, *25*(6), 751–762.

Chinchilla, A., & Garcia, M. (2017). Social entrepreneurship intention: Mindfulness towards a duality of objectives. *Humanistic Management Journal*, *1*(2), 205–214.

Clift, S., & Camic, P. (2016). *Oxford textbook of creative arts, health, and well-being: International perspectives on practice, policy and research*. New York: Oxford University Press.

Dessein, J., Soini, K., Fairclough, G., & Horlings, L. G. (Eds.). (2015). *Culture in, for and as sustainable development: Conclusions from the COST Action IS1007 Investigating Cultural Sustainability*. Jyväskylä, FI: University of Jyväskylä.

Dierksmeier, C. (2016). What is 'humanistic'about humanistic management? *Humanistic Management Journal*, *1*(1), 9–32.

Duxbury, N., Campbell, H., & Keurvorst, E. (2011). Developing and revitalizing rural communities through arts and culture. *Small Cities Imprint*, *3*(1), 111–122.

Duxbury, N., Hosagrahar, J., & Pascual, J. (2016). *Why must culture be at the heart of sustainable urban development?* [Policy paper]. Barcelona: United Cities and Local Governments.

Duxbury, N., Kastenholz, E., & Cunha, C. (2019). Co-producing cultural heritage experiences through creative tourism. In W. Gronau, R. Bonadei, E. Kastenholz, & A. Pashkevich (Eds.), *E-CuL-Tours: Enhancing networks in heritage tourism* (pp. 189–206). Rome, IT: tab edizioni.

Duxbury, N., & Richards, G. (2019). Towards a research agenda for creative tourism: Developments, diversity, and dynamics. In N. Duxbury & G. Richards (Eds.), *A research agenda for creative tourism* (pp. 1–14). Cheltenham, UK: Edward Elgar Publishing.

Duxbury, N., Silva, S., & Castro, T. V. de. (2019). Creative tourism development in small cities and rural areas in Portugal: Insights from start-up activities. In D. A. Jelinčić & Y. Mansfeld (Eds.), *Creating and managing experiences in cultural tourism* (pp. 291–304). Singapore: World Scientific Publishing.

Kabadayi, S., Alkire, L., Broad, G. M., Livne-Tarandach, R., Wasieleski, D., & Marie Puente, A. (2019). Humanistic management of Social Innovation in Service (SIS): An interdisciplinary framework. *Humanistic Management Journal*, *4*(2), 159–185. doi:10.1007/s41463-019-00063-9

Kagan, S. (2011). *Art and sustainablity: Connecting patterns for a culture of complexity*. Bielefeld, Germany: Transcript Verlag.

Kallis, S. (2014). *Common threads: Weaving community through collaborative eco-art*. Gabriola Island, BC: New Society Publishers.

Landry, C. (2010). Experiencing imagination: Travel as a creative trigger. In R. Wurzburger et al. (Eds.), *Creative tourism, a global conversation: How to provide unique creative experiences for travelers worldwide* (pp. 33–42). Santa Fe, NM: Sunstone Press.

Langer, E. J. (2005). *On becoming an artist: Reinventing yourself through mindful creativity*. New York: Ballantine Books.

Laszlo, C. (2019). Strengthening humanistic management. *Humanistic Management Journal*, *4*, 85–94.

Little, B. R., Salmela-Aro, K., & Phillips, S. D. (2014). *Personal project pursuit: Goals, action, and human flourishing*. New York: Psychology Press.

Luckman, S. (2012). *Locating cultural work: The politics and poetics of rural, regional and remote creativity*. London: Palgrave Macmillan.

Melé, D. (2016). Understanding humanistic management. *Humanistic Management Journal*, *1*(1), 33–55. doi:10.1007/s41463-016-0011-5

Miller, A. (2017). Creative geographies of ceramic artists: Knowledges and experiences of landscape, practices of art and skill. *Social and Cultural Geography, 18*(2), 245–267.

Natura. (2009). *Natura annual report 2008*. São Paulo: Natura Brasil.

Nussbaum, M. (2010). *Not for profit: Why democracy needs the humanities*. Princeton, NJ: Princeton University Press.

Pirson, M. (2018). Reclaiming our humanity: A cornerstone for better management. *Humanist Management Journal*, *2*(2), 103–107.

Riccardi, V., & Ferreira, V. (Eds.). (2019). *Creative responses to sustainability: Cultural initiatives engaging with social and environmental issues. Portugal Guide*. Singapore: Asia-Europe Foundation (ASEF).

Richards, G. (2011). Creativity and tourism: The state of the art. *Annals of Tourism Research*, *38*(4), 1225–1253. doi:10.1016/j.annals.2011.07.008

Rodríguez-Lluesma, C., Davila, A., & Elvira, M. M. (2014). Humanistic leadership as a value-infused dialogue of global leaders and local stakeholders. In N. C. Lupton & M. Pirson (Eds.), *Humanistic perspectives on international business and management* (pp. 81–91). London: Palgrave Macmillan.

Ross, S. L. (2010). Transformative travel: An enjoyable way to foster radical change. *ReVision*, *32*(1), 54–61.

Ryff, C. D. (2014). Psychological well-being revisited: Advances in the science and practice of eudaimonia. *Psychotherapy and Psychosomatics*, *83*(1), 10–28. doi:10.1159/000353263

Ryff, C. D., & Singer, B. (2008). Know thyself and become what you are: A eudaimonic approach to psychological well-being. *Journal of Happiness Studies*, *9*, 13–39.

Steckler, E. L., & Waddock, S. (2018). Self-sustaining practices of successful social change agents: A retreats framework for supporting transformational change. *Humanistic Management Journal*, *2*(2), 171–198.

UNWTO. (2016). *Global report on the transformative power of tourism: A paradigm shift towards a more responsible traveller*. Madrid, Spain: UNWTO and Institute for Tourism, Zagreb, Croatia.

Wolf, I. D., Ainsworth, G. B., & Crowley, J. (2017). Transformative travel as a sustainable market niche for protected areas: A new development, marketing and conservation model. *Journal of Sustainable Tourism*, *25*(11), 1650–1673. doi:10.1080/09669582.2017.1302454

Wright, P. R., & Pascoe, R. (2015). Eudaimonia and creativity: The art of human flourishing. *Cambridge Journal of Education*, *45*(3), 295–306. doi:10.1080/0305764X.2013.855172

8 A Bloody Past With a Bright Future

Dark Tourism and Humanistic Management at Missouri State Penitentiary

Daniel P. Bumblauskas and Salil S. Kalghatgi

1. Introduction and Research Motivation

The Missouri State Penitentiary (herein referred to as the "MSP" or "prison") is a historic landmark in Jefferson City, Missouri (United States). Operated until 2004, this now de-commissioned prison is owned by the State of Missouri and is being utilized for various types of tours and leased by the Jefferson City Convention and Visitors Bureau. One of the primary research questions explored in this chapter is whether renovating the prison in Jefferson City, to prevent aesthetic pollution, would be a worthwhile investment based on forecasted tourist traffic and revenue-generating potential. Beyond this, the overall humanistic impact of the prison is considered, specifically the educational potential at hand in giving the public a rare opportunity to empathize with the incarcerated. If the MSP connects tourists and stakeholders with an otherwise disenfranchised group, then it will have achieved more than maintaining a historical or cultural landmark; it will serve as a catalyst to also rehabilitate and preserve the dignity of fellow members in our local communities.

The condition and aesthetic appearance of the prison has historically caused some opposition to the prison's existence, during both its former active prison days and now as a place of tourism. While much of the opposition seems to have subsided in recent years, some of the environmental and aesthetic concerns of the surrounding businesses and community include structural stability, hazardous materials, and general appearance. A literature review of aesthetic pollution, prison tour operators, and the dark tourism theoretical construct has shown that this is a rapidly growing area of interest in tourism management research as well as a protection of dignity for the common good—a topic we explore through the concept of 'sites of conscience' at another prison turned tourist facility.

This chapter details some of the lease/ownership history and arrangement between the city and state and documents challenges faced related to sustainable operations and humanistic management at the MSP dark tourism site. A summary and analysis of the tours offered, visitor data, and revenue data are provided along with an assessment of the MSP compared with benchmarked peer institutions. Finally, conclusions and recommendations are given for the continued operation of the MSP to maximize its financial and social value.

2. Literature Review

Kaplan (2003) is one of the first to bring the fields of aesthetic pollution and dark tourism into alignment with his discussion of holocaust sites; Hartmann (2014) also details the holocaust sites written extensively about by Ashworth (1996, 2002). Baddeley (2004) incorporates the importance of aesthetics of tourist sites in terms of revenue generation in Thailand which Hartmann (2014) refers to broadly as "landscapes, builtscapes, workscapes, technoscapes, and peoplescapes."[1] The MSP is located in downtown Jefferson City and has had a major impact on the community economically and as it relates to visual or aesthetic pollution in the downtown area. Visual and aesthetic pollution have been studied from an ontological perspective (Douglas, 1966; Douglas, 2003) and in numerous applications such as water and beachfront photography (Tudor & Williams, 2003), empirically for an electrical power generation plant (Randall, Ives & Eastman, 1974), and in general practice (Holden, 2008).

Foley and Lennon (1995, 1996, 1997) were some of the first researchers to introduce the terminology "dark tourism," meaning "tourism associated with sites of death, disaster, and depravity" (Lennon & Foley, 1999, p. 46). The dark tourism theory and field has since been further expanded upon (Lennon & Foley, 1999; Lennon & Foley, 2000; Cannon-Brookes, 2001; Smith, 2002; Tarlow, 2007; Wight & Lennon, 2007) and has been studied for sites of mass genocide in Cambodia (Isaac & Çakmak, 2016), holocaust and Nazi socialist sites (Aschauer, Weichbold, Foidl and Drecoll, 2017; Ashworth, 1996; Ashworth, 2002; Jacques, 2006), and celebrity grave sites, such as that of Marilyn Monroe (Baidwan, 2015). Prisons have been studied by Strange and Kempa (2003) and Hedges (2014) in the context of dark tourism for Alcatraz (California, United States) and Robben Island (South Africa). Wight and Lennon (2007) and Kang, Scott, Lee and Ballantyne (2012) explore cases for a dark tourist attraction(s) from "natural and man-made disasters or atrocities" (Kang et al., 2012, 257). Stone and Sharpley (2008) note the need for research related to the demand for dark tourism, and Strange and Kempa (2003) note the need for more research on marketing, interviews with prison tour operators, and reviewing visitor expectations as conducted in this research on the MSP. Yuill (2004) reviews dark tourism in more detail based on the work of Dunlap (2001) for the Sing Sing Prison (New York,

United States) and drawing comparisons to the Pompeii (Italy) natural disaster tourist site.

Lennon (2017) recounts the history of dark tourism, as detailed further in the literature review, which includes defining the field (Seaton, 1996). Further, Stone (2006) calls the field "eclectic and theoretically fragile," arguing that there are "shades of darkness" for such sites (as well as considering demand and supply alignment). Stone (2006) as well as Bowman and Pezzullo (2009) and Hartmann (2014) discuss macabre tourist sites and question the field and terminology of "dark tourism" entirely. Hartmann (2014) provides a good review of the history of the field, with many of the early researchers located in Scotland and cascading from there. While this discussion is still on-going in the scholarly literature among academics, these sites are still operating and growing in interest. Many of the key foundational researchers in dark tourism published the *Palgrave Handbook of Dark Tourism Studies* (Stone, Hartmann, Seaton, Sharpley & White, 2018).

Figure 8.1 is an aerial photo of the MSP site (Missouri State Penitentiary Redevelopment Commission, 2002). Figure 8.2 shows the aerial layout of the prison, including the cell blocks, chapel, factory, gas chamber, etc., on the prison grounds (circled). The surrounding real estate and property have suffered over the years from aesthetic pollution to the area.

The MSP can be classified as a tourist attraction or site given its status as a historic monument and cultural heritage site. Figure 8.3 (2011) and

Figure 8.1 Missouri State Penitentiary site

Source: Missouri State Penitentiary Redevelopment Commission

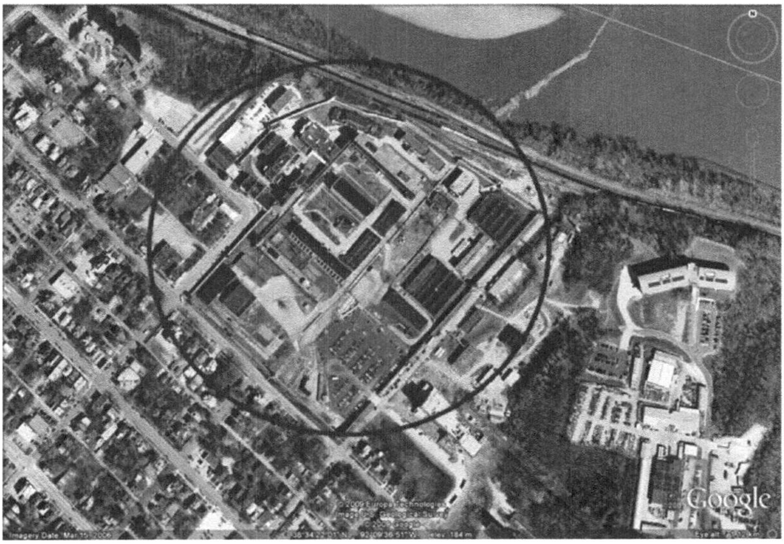

Figure 8.2 Aerial layout of Missouri State Penitentiary

Source: Google Map

Figure 8.3 MSP informational and marketing leaflet (2011)

Source: Jefferson City Convention and Visitor Bureau

Figure 8.4 MSP tri-fold leaflet (2016)

Source: Jefferson City Convention and Visitors Bureau

Figure 8.4 (2016) contain sample informational and marketing tri-fold leaflets with an overview of the wide variety of MSP's historically significant events and tours that have been offered over the years. The role of tour guides at historical sites has been considered by others (Reisinger & Steiner, 2006) as well as creating a typography for the types of tours offered, specifically for wildlife tours (Curtin & Wilkes, 2005). Poria, Butler and Airey (2003) looked at the impact of tourism on visitation to religious historical sites, and some commonalities exist between their work and that of historical sites such as the MSP. The MSP was the first prison west of the Mississippi River and in 1967 was deemed by *Time* magazine as the "bloodiest 47 acres in America" (Missouri Department of Economic Development, 2020) and housed many famous inmates (e.g. Sonny Liston, 'Pretty Boy' Floyd, etc.) before it closed in 2004. The prison has become a major tourist destination for the state of Missouri. However, there has historically been a faction of residents who have been interested in demolishing the prison because of its condition and aesthetic pollution to the downtown Jefferson City area.

This work builds upon prison re-development in terms of cultivating tourism rather than alternative or private management of prisons.

Public-private development and re-development activities in Australia (Duffield, 2005), inmate re-development (Salomone, 2004a; Salomone, 2004b), and the development of new correctional facilities and their impacts (Bonds, 2013) have been documented. A new prison was constructed in the surrounding area outside the city center.

3. Methodology

As we see from the literature review that the field of dark tourism is relatively new (Lennon & Foley, 2000; Hartmann, 2014), research such as this continues to build upon the dark tourism construct. To establish the visitor and revenue potential, a comparative analysis was conducted to compare the MSP with other prisons, such as Alcatraz in California. This revenue stream is pivotal to the preservation of the structure and allows the public to continue learning about the penal system. Alcatraz has become a major tourist destination for the San Francisco Bay area, and each location was visited by the author(s) on multiple occasions. The MSP operated from 1836 to 2004 (Jefferson City Convention and Visitors Bureau, [2011] 2020) and is 100 years older than Alcatraz. In working with the Jefferson City Convention and Visitors Bureau to investigate the current status of the prison, it was decided that benchmarking other peer institutions would be beneficial. Our research goal was comparing the data to the number of tourists and revenues generated by peer institutions and forecasting the number of visitors and revenues the MSP could expect in a given period. The goal of this analysis was to aid in justifying or refuting whether renovations at the MSP should be conducted to reduce aesthetic pollution by increasing the number of tours to fund necessary repairs.

An original motivation for this study was that the municipality of Jefferson City had been reviewing plans to build a convention center at the MSP site, and some of these preliminary plans called for the demolition or redevelopment of the prison and residential housing in the surrounding area (KOMU News, 2007, 2010). One of the primary challenges encountered in conducting this study was locating the data and the information needed from the state of Missouri and those from peer institutions. With the assistance of Diane Gillespie and Steve Picker, executive director and former executive director of the Jefferson City Convention and Visitors Bureau, respectively, contact information was obtained for peer sites, and eventually public reports and resources were located which describe the impact on the ambient environment in terms of aesthetic pollution, tourist traffic, and revenues from each prison. The Jefferson City penitentiary has only been offering tours since 2009 (Uhlenbrock, 2009), so learning from the lessons of Alcatraz and others has been extremely beneficial to the MSP's operations.

In the case of the MSP, Jefferson City and the state of Missouri must be careful in considering the aesthetic pollution and revenue-generation model. As noted by Holden (2008), "Often tourism development is based upon maximizing profits whilst ignoring aesthetic concerns (Holden, 2008)." Jefferson City must generate sufficient revenue to cover the restoration costs. As such, the prioritization of restoration activities must be accounted for (e.g. a leaking roof was replaced to prevent more extensive damage). One theory is that successful restoration requires a balance between and/or shift from extrinsic to intrinsic motivators over time as we persuade communities, politicians, business people, and other stakeholders, of the need to perform such work—for example, the use of Herzberg's Hygiene Theory (Heizer & Render, 2010) for satisfaction and a shift in focus toward educating the aforementioned stakeholders from extrinsic measures (revenues, profits, etc.) toward humanistic management measures (e.g., sites of conscience, rehabilitation, reincarceration rates, etc.).

The MSP also has great educational value to offer. Most of the American and global public do not understand what it means to be incarcerated, but the authors believe that having access to a defunct prison will give tourists the opportunity to empathize with the very difficult life many prisoners lead and the socio-economic context of most prisoners, along with the fallout among poor and minority communities who tend to have an absence of male leaders. Some feel uncomfortable raising these issues and do not think it aligns with a positive experience. To better understand how social education affects a tour offering, we lean on the data of 251 surveys provided by the Eastern State Penitentiary for their exhibit, *Prisons Today: Questions in the Age of Mass Incarceration* (Kelley, 2017; Tinker, 2016). This exhibit was partially funded by a United States federal Pew Grant and explored educating tourists on the phenomenon of disproportionately high incarceration rates among the poor, people of color, and disenfranchised. The grant's purpose is to "inform organizations on how to present current and provocative issues" and to reduce recidivism. Other examples include teaching and education in prisons (Ferry, 2020); humanistic management and empathy considerations are also important at the MSP—that is, the power of sites of conscience in aiding and rehabilitating the incarcerated. In the United States, this is also a bridge to current governmental policy, for example, the First Step Act of 2018 (Public Law 115–391, 2018).

Jefferson City is the state capital of Missouri and is located in the center of Missouri between St. Louis, to the east, and Kansas City, to the west. The prison facility is owned by the state of Missouri, which limits Jefferson City's ability to fully control operations and make such prioritization decisions at the MSP. The city and the state had utilized lease agreements up to 2011 and had been evaluating longer-term lease options, which led to a 15-year use agreement with two additional five-year option agreements (Gillespie, 2016).

The state of Missouri established the Missouri State Penitentiary Redevelopment Commission (MSPRC) to plan for the sustainable operation of the MSP. Figure 8.5 provides a sample map of one of the redevelopment

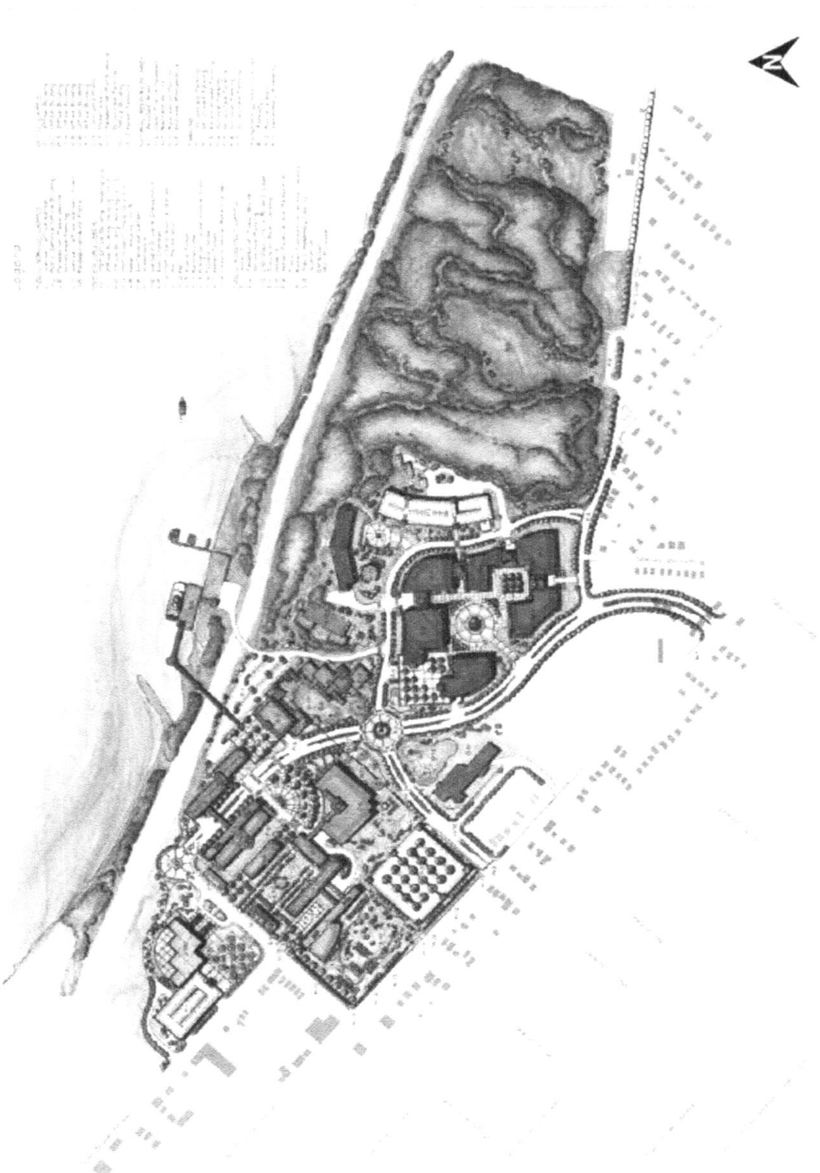

Figure 8.5 State of Missouri MSP Redevelopment Commission design guidelines

Source: Missouri State Penitentiary Redevelopment Commission

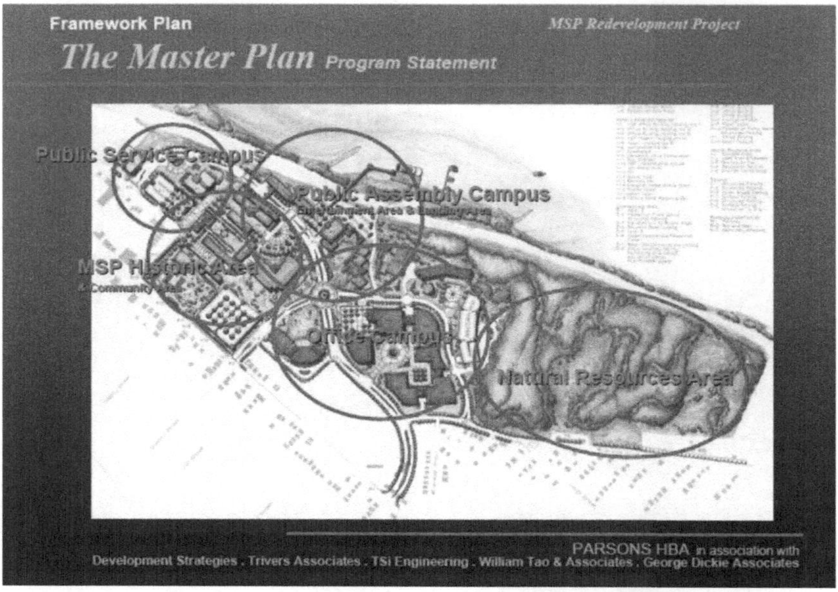

Figure 8.6 Framework plan for the MSP property

Source: Missouri State Penitentiary Redevelopment Commission

guidelines issued by the state of Missouri (State of Missouri Division of Facilities Management, 2006).

The site itself represents a fairly large portion of the city's official land mass at approximately 0.5%. However, only five of those 137 acres are considered to be the "historic district" portion of the prison. Some of the environmental and aesthetic pollution concerns on the property included structural stability of buildings, hazardous materials such as asbestos and lead-based paints, boarded-up entrances and windows, and general maintenance and upkeep such as lawn care and snow removal. The MSPRC Master/Framework Plan (Missouri State Penitentiary Redevelopment Commission, 2002) outlines some of these topics by prioritizing buildings based on historical value, analyzing structural conditions, providing architectural summaries, reviewing engineering considerations and challenges, and initiating environmental investigations. Figure 8.6 provides a snapshot from the MSPRC Master/Framework Plan conceptual layout for the MSP property.

The MSPRC studied the existing structures and analyzed some of the environmental impacts and deemed that there are minimal hazardous material concerns at the site. Some of the challenges faced by Jefferson City in operating the MSP include (Picker, 2011):

- Funding to stabilize the structures (improve aesthetics; e.g. new roofs, windows, etc.)
- Redevelopment activities including the historic district and private/public partnerships
- Prioritization of ideas from various organizations such as the city, chamber of commerce, state of Missouri, and the MSPRC
- Staffing MSP during normal business hours (e.g. 9 AM—5 PM) which seems to be cost prohibitive in terms of salaries, insurance, and utilities (e.g. lighting, trash, electric, water, etc.)

Revenue generation in this case leads to acquire helps the multiple stakeholders, including the public, to bond and comprehend the social complexities at hand in today's criminal justice system (Pirson & Lawrence, 2009). Before a comprehensive sustainability strategy could be deployed for the MSP, the state and city needed to improve the condition of the current facilities to acceptable levels. The primary historical revenue-generation method is tours. Table 8.1 shows the various MSP tour options and includes additional details on each tour type.

Table 8.1 Missouri State Penitentiary tour options

Tour type	*Cost*	*Description*	*Other notes*
Two-hour history tour	$12 per person	Regular tour	—
Four-hour in-depth tour	$25–35 per person	Discover tour	—
Twilight tour	$17 per person	6–11 PM	Visitors are given a lantern to investigate
Specialty tour	$17 per person	Focus on special events and people	"Pretty Boy" Floyd MSP escape attempts
Ghost tour	$25 per person	Unusual occurrences	Use of activity-finding devices in two of the housing units and gas chamber
Overnight	$95 per person	Up all night, no sleeping accommodations	'A' hall and gas chamber access

Source: Authors' own elaboration based on Picker (2011).

Table 8.2 and Figure 8.7 show the visitor and revenue data for the MSP from 2009 to 2017, and ticket sales data from 2009 to 2017, respectively. The average annual visitor growth rate from 2009 to 2017 has been 61%, and the average annual revenue growth rate from 2009 to 2017 has been 59%. Figure 8.8 shows the revenue growth per guest over that period which increased from $12.66 (2009) to $17.43 (2015).

Projected revenue for 2013 was estimated from the actual revenue per guest from that year. Visitor counts from 2016 and 2017 were also estimated based on actual ticket revenues. Diane Gillespie (2016) noted that in 2013, the MSP was forced to close early in September 2013

Table 8.2 MSP visitor and revenue data

Year	*Ticket revenue ($)*	*Visitors*	*Revenue per visitor ($)*	*Annual visitor growth*	*Annual revenue growth*	*Revenue growth since inception*
2009	41,650	3,290	12.66	—	—	—
2010	176,000	12,000	14.67	265%	323%	323%
2011	263,000	17,203	15.29	43%	49%	531%
2012	275,471	19,121	14.41	11%	5%	561%
2013	210,000	14,581	14.40	-24%	-24%	404%
2014	301,409	22,605	13.33	55%	44%	624%
2015	452,104	25,945	17.43	15%	50%	985%
2016	536,768	36,765	14.60	42%	19%	1189%
2017	553,904	37,939[1]	14.60	3%	3%	1230%

Source: Authors' own elaboration of Picker (2011) and Gillespie (2016), executive directors, Jefferson City Convention and Visitors Bureau ([2011] 2020)

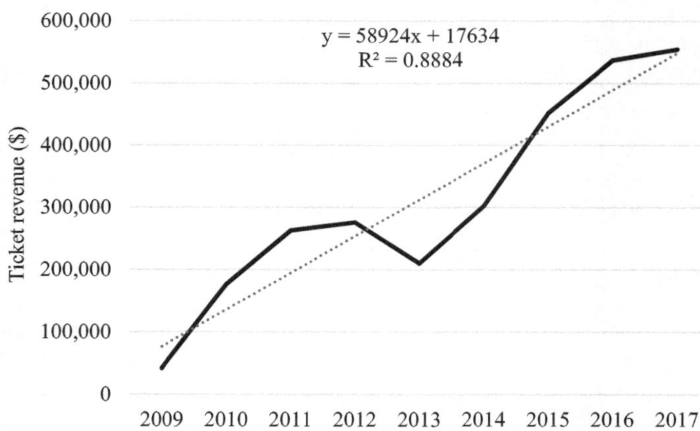

Figure 8.7 MSP ticket sales from 2009 to 2017

Source: Author's own elaboration

Figure 8.8 MSP revenue per guest from 2009 to 2015

Source: Authors' own elaboration

because of unexpected repair needs. This early closure affected the number of visitors as site personnel were not able to offer tours and services during their normal operating period of March 1 to November 30. Projects funded during this repair period included window replacements, mold eradication, and lead-based paint mitigation. In addition, MSP had to cancel 3,000 reservations which impacted local hotels, restaurants, etc., and had economic ripple effects throughout the community (Gillespie, 2016).

4. Benchmarking

Jefferson City's population in 2010 was approximately 41,297 (City-Data, 2010), and the number of visitors to the prison during 2011, following closure for the season, was 17,203 (Picker, 2011); at the end of 2015 it was 25,945 (Gillespie, 2016). In 2004, when the MSP was opened to the public for the first time, 20,000 visitors attended the grand opening event (Cabbage & Bureau, 2011). Alcatraz sees 1.4 million annual visitors (Golden Gate National Parks Conservancy 2011); the total population in San Francisco was 808,976 in 2011 (U.S. Census Bureau and Google Data, 2008). That gives Alcatraz a visitor-to-resident ratio of 1.73; the ratio for the MSP was 0.42 for 2011, as shown in Table 7.3. Peer prisons include (Picker, 2011):

- West Virginia Penitentiary (www.wvpentours.com/)
    - Moundsville, West Virginia; Population: 9,054

- Ohio State Reformatory (www.mrps.org/)
  - Owned by the Mansfield Reformatory Preservation Society
  - Mansfield, Ohio; Population: 49,346
- Eastern State Penitentiary (www.easternstate.org/)
  - Philadelphia, Pennsylvania; Population: 1,447,395

Eastern State Penitentiary in Philadelphia is a "preserved ruin" (Visions, 2019, p. 22) and as of 2011 was attracting in excess of 250,000 visitors annually in 2011 (Elk, 2011), but it has a much smaller visitor-to-resident ratio of 0.17 given the large population of the greater Philadelphia metropolitan area. The number of visitors has since grown at an impressive average annual rate of 13% from 2002 to 2018, for a total visitor count in 2018 of 418,218 (Frankhouser & Kelley, 2019). In addition, the city of Philadelphia also reports 37.4 million visitors annually (Greater Philadelphia Tourism Marketing Corporation, 2011). Other prisons used for comparison in Table 7.3 include Alcatraz in San Francisco. Data were not readily available from each peer institution; however, data for those with available information are shown in Table 8.3.

While perhaps not a direct comparison, Alcatraz has a fascinating history which closely aligns with the MSP in scale. According to Tom Kiely, executive vice president of tourism for San Francisco Travel, Alcatraz is used as an instrumental part of the marketing strategy for the San Francisco area. Kiely said that "San Francisco received approximately 16M visitors in 2010, generating $8B in revenue for the city, supporting 70K jobs and contributing $500M in tax revenue to the city's general fund. It

Table 8.3 Prison comparisons by population, visitors per year, and tourists per year (2011)

Location	*Population*	*City visitors per year*	*Prison site tourists per year (ratio)*	*Source*
Jefferson City, Missouri (MSP)	41,297	N/A	17,203 (0.42)	U.S. Census (2008) Picker (2011)
San Francisco, California (Alcatraz)	808,976	16 million	1.4 M (1.73)	US Census Kiely (2011) Phelps (2011) California state parks
Philadelphia, Pennsylvania (ESP)	1,447,395	37.4 million	250,000 (0.17)	Elk (2011) City of Philadelphia

Source: Authors' own elaboration based on listed sources in table.

is our number one industry" (Kiely, 2011). One constraint on Alcatraz is that US federal guidelines only allow 3,000 visitors per day during peak portions of the summer which limits the total number of guests allowed to visit the site. Nicki Phelps from the Parks Conservancy suggested that the number of visitors per year could double if it was allowed (Phelps, 2011). Site locations for Alcatraz and the MSP are both highly desirable, with Alcatraz being located in San Francisco Bay and the MSP in downtown Jefferson City, situated between the major metropolitan areas of St. Louis and Kansas City, along the Missouri River. Akin to the large growth in development in the San Francisco area, the MSP is also bringing in developers and sending requests for quotation (RFQ's) for varied residential and consumer developments on 31 acres of land (News Tribune, 2019a).

One challenge in attaining information from other peer institutions (e.g. West Virginia Penitentiary and Ohio State Reformatory) is the lack of resources at smaller historical sites and the seasonal closure of the facilities during the winter. For example, Jefferson City and Marshall County (West Virginia Penitentiary) do not have the types of data collection resources which are available to the city of Philadelphia and San Francisco. Most of the facilities close entirely or have limited hours during the winter. The data from peer institutions illustrate the market potential for the MSP. With hundreds of thousands, even millions of visitors touring some of these facilities, there is potential to capture a larger market for prison tours.

5. Results and Recommendations

One of the primary decisions to be made is to what extent the prison should be restored and further developed for historical tours and/or commercial use. Because Jefferson City is not "land-locked," and therefore not in need of land to be developed, much of the opposition to the MSP continuing to offer tours with a facility in less than ideal aesthetic condition has been alleviated. There is a new $80 million federal courthouse directly across the street from the MSP and that Jefferson City officials anticipated some complaints about the exterior of the MSP once that facility was fully operational. As such, some parts of the exterior renovation were expected to be ranked higher in priority because of political and community pressures. It was also anticipated that redevelopment would increase surrounding property values, primarily residential properties directly bordering the MSP. Many of the residential properties in the area have been abandoned, so there is the potential for indirect aesthetic improvements to surrounding neighborhoods.

Data from the period 2015–2017 have shown a slowdown in the pace of revenue growth at the MSP. However, there is consistent evidence that visitor and revenues are still trending upward both at the MSP and at benchmark institutions. Benchmarking against peer prison-tourism institutions is difficult because of the relatively few players in the space and the limited resources available at the smaller-sized operations. Therefore, the focus was on more granular financial and visitor data to find strong operating fundamentals at the MSP. The greatest threat since MSP's opening has been unexpected capital improvement expenses which include 1) the closure in 2013 for nearly half the season as a result of unforeseen repairs and 2) a tornado that struck Jefferson City on May 22, 2019, halting tours until March 2020 and resulting in an estimated $4.2 million in repairs (News Tribune, 2019b). This suggests that when managing a historical site, one should expect greater expense variances and heavily scrutinize the short-term forecast accuracy of a (linear) trend (Figure 8.9).

For the state of Missouri and Jefferson City to see consistent value in light of the sustainability challenges when addressing aesthetic pollution, there must be a material benefit to the state's population. The educational component of a dark tourism facility gives the tourist both an enjoyable experience and an opportunity to learn about the criminal justice system. The *Prisons Today* exhibit at Eastern State Penitentiary "was a success (rated 6.2 out of a possible 7), . . . almost all visitors (94%) understood some, or all, of the main message of the exhibit" overall, demonstrating that it is possible to raise social commentary while still creating a positive experience (Tinker, 2016). The operational success comes from having tours that both generate revenue and educate visitors about major issues in America's criminal justice system. Interviews with staff at the

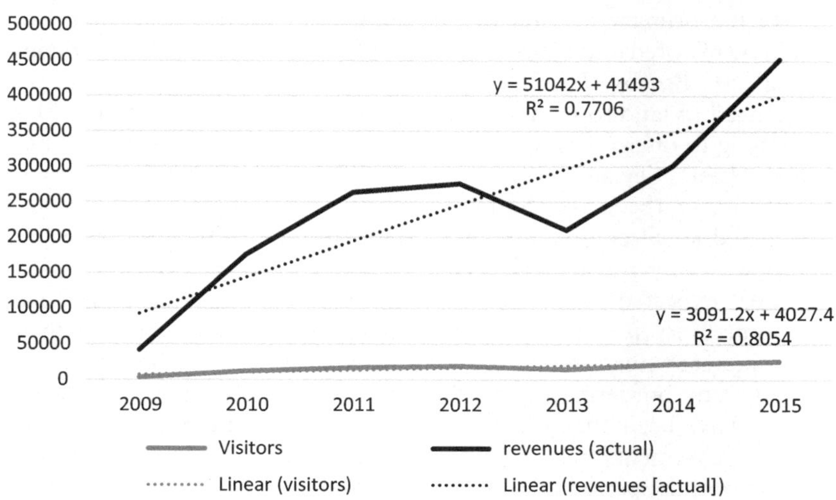

Figure 8.9 MSP visitor and revenue data with linear trend forecasts (2009–2015)

Source: Authors' own elaboration

Eastern State Penitentiary Historic Site reinforced their commitment to education from the organizational leadership as paramount to becoming a 'Site of Conscience', a global network of historic sites, museums and memory initiatives connecting past struggles to today's movements for human rights and transitional justice (www.sitesofconscience.org/en/who-we-are/about-us/).

Recommendations for the MSP to increase value include the following:

- Utilize a benefit analysis to prioritize redevelopment projects and forecast for infrequent but significant repairs adjusted for heritage maintenance.
- Join or launch a network of historical prison tour operators to share best practices, create industry benchmarks, and leverage shared resources whenever possible.
- Secure additional funding opportunities and create a positive experience by raising social awareness of national criminal justice issues.
- Partner with educational centers, such as Lincoln University, a historically black college in Jefferson City, and invite them to become stakeholders in exhibits.
- Lengthen the operating schedule to offer year-round tours.
- Provide shuttle services from St. Louis and Kansas City (e.g. via Amtrak).
- Develop office space for state and local government employees (State of Missouri, Division of Design & Construction, 2002).
- Optimize advertising campaigns' Return on Investment (ROI) through analytics.

6. Conclusion and Future Research

There is a great deal of opportunity and excitement surrounding the continued redevelopment, sustainability management practices, "dark tourism," revenue potential, and humanistic management at the MSP. One observation about historic prisons in general is that the supply chain support by third parties seems to be more limited than other tourism industries. Many of the prison sites operate independently, and the support network in the form of associations is less developed than what is observed in other sectors such as the American Hotel and Lodging Association (www.ahla.com/). As operations continue to expand, additional peer benchmarking would be beneficial to all parties in the industry. An ideal future state may involve initiatives such as group purchasing organizations (GPOs) where smaller independent sites can leverage shared economies of scale for better supplier results. With a relatively small number of prison tourism sites, future partnerships could likely include international peers such as the Kilmainham Gaol in Dublin, Ireland (www.heritageireland.ie/en/dublin/kilmainhamgaol/), which is now one of the top tourist attractions in all of Ireland, with tickets increasingly difficult to obtain.

Our analysis of MSP's financials suggest that maintenance expenses and advertising spend are the two most significant opportunities to drive financial success, and the authors believe they warrant further inspection, likely including improved operational forecast modeling as well as optimizing advertising ROI. This would take careful coordination between the necessary cross-functional teams and may be a good opportunity to partner with peer institutions, research foundations, or consultants.

Dark tourist sites like prisons can also bring greater social value to the community by leveraging a location's heritage as an opportunity to create a dialogue. It is reassuring to see that efforts promoting institutional changes have been met with positive feedback. There are 2.2 million American citizens in prison or jail, and it can be difficult for many in the public can empathize with a sequestered population. The impact of educating the regional public also has significant value because when the MSP closed its prison operations, a new prison was opened nearby. There are also programs being offered by organizations such as Iowa Prison Industries that are targeting job skills and training in areas such as Lean Six Sigma to provide value to individuals as they are released back into society (as well as businesses and organizations in the communities in which they reside). This gives the MSP the rare opportunity to create a dialogue about the impact of social justice within the greater state and regional areas. The alignment of dark tourist sites, such as historic prisons, can be used to empathize with incarcerated individuals, thereby achieving a major stepping stone toward a more just society.

Note

1. Note that Hartmann (2014), via personal communications with Ashworth, proposes there are no such dark tourist sites, only dark tourists. See also Ashworth and Isaac (2015).

References

Aschauer, W., Weichbold, M., Foidl, M., & Drecoll, A. (2017). Obersalzberg as a realm of experience on the quality of visitors' experiences at National Socialist places of remembrance. *Worldwide Hospitality and Tourism Themes*, *9*(2), 158–174.

Ashworth, G. J. (1996). Holocaust tourism and Jewish culture: The lessons of Krakow-Kazimierz. In M. Robinson, N. Evans, & P. Callaghan (Eds.), *Tourism and cultural change* (pp. 1–12). Newcastle: Centre for Travel and Tourism.

Ashworth, G. J. (2002). Holocaust tourism: The experience of Krakow-Kazimierz. *International Research in Geographical and Environmental Education*, *11*(4), 363–367. doi:10.1080/10382040208667504

Ashworth, G. J., & Isaac, R. K. (2015). Have we illuminated the dark? Shifting perspectives on 'dark' tourism. *Tourism Recreation Research*, *40*(3), 316–325.

Baddeley, M. C. (2004). Are tourists willing to pay for aesthetic quality? An empirical assessment from Krabi Province, Thailand. *Tourism Economics*, *10*(1), 45–61.

Baidwan, S. (2015). *Six feet from fame: Exploring dark tourism motivation at Marilyn Monroe's grave* (Dissertation), University of Nevada, Las Vegas.

Bowman, M. S., & Pezzullo, P. C. (2009). What's so 'dark' about 'dark tourism'? Death, tours, and performance. *Tourist Studies*, *9*(3), 187–202.

Bonds, A. (2013). Economic development, racialization, and privilege: "Yes in my backyard" prison politics and the reinvention of Madras, Oregon. *Annals of the Association of American Geographers*, *103*(6), 1389–1405.

Cabbage, C., & Bureau, J. C. (Directors). (2011). Missouri State Penitentiary [Motion Picture].

Cannon-Brookes, P. (2001). Dark tourism: The attraction of death and disaster-By John Lennon and Malcom Foley. *Museum Management and Curatorship*, *19*(2), 221–223.

City-Data. (2010). City-Data Jefferson City, MO. Retrieved February 20, 2020, from www.city-data.com/housing/houses-Jefferson-City-Missouri.html

Curtin, S., & Wilkes, K. (2005). British wildlife tourism operators: Current issues and typologies. *Current Issues in Tourism*, *8*(6), 455–478.

Douglas, M. (1966, 2003). *Purity and danger: An analysis of the concepts of pollution and taboo*. New York: Routledge.

Duffield, C. F. (2005). PPPs in Australia. Public private partnerships: Opportunities and challenges. *Proceedings of the CICID*, *22*, 5–14.

Dunlap, D. (2001, May 2). Making a museum where thousands did their time. *New York Times*. Retrieved February 20, 2020, from www.nytimes.com/2001/05/02/arts/02DUNL.html

Elk, S. (2011, December 2). Interview with President and CEO, Eastern State Penitentiary Historic Site, Inc. (D. Bumblauskas, Interviewer).

Ferry, J. (2020). Why I go to prison. *Academe*, *106*(1). Retrieved February 10, 2020, from www.aaup.org/article/why-i-go-prison

Foley, M., & Lennon, J. J. (1995). JFK and cultural tourism. In *First annual conference for popular culture*. Manchester: Manchester Metropolitan University.

Foley, M., & Lennon, J. J. (1996). JFK and dark tourism: A fascination with assassination. *International Journal of Heritage Studies*, *2*(4), 198–211.

Foley, M., & Lennon, J. J. (1997). Dark tourism—an ethical dilemma. *Hospitality, Tourism and Leisure Management: Issues in Strategy and Culture*, 153–164.

Frankhouser, N., & Kelley, S. (2019). Eastern state penitentiary. Email correspondence dated March 11, 2019.

Gillespie, D. (2016, March 24). Executive Director, Jefferson City Convention and Visitors Bureau. (D. Bumblauskas and S. Kalghatgi, Interviewers).

Golden Gate National Parks Conservancy. (2011). *Alcatraz*. Retrieved December 12, 2011, from http://www.parksconservancy.org/visit/alcatraz/alcatraz-tours.html

Greater Philadelphia Tourism Marketing Corporation. (2011). *Research and publications*. Philadelphia and the Countryside. Retrieved February 20, 2020, from www.visitphilly.com/research/

Hartmann, R. (2014). Dark tourism, thanatourism, and dissonance in heritage tourism management: New directions in contemporary tourism research. *Journal of Heritage Tourism*, *9*(2), 166–182.
Hedges, C. (2014). Alcatraz: A prison as Disneyland. *TruthDig*. Retrieved February 20, 2020, from www.truthdig.com/articles/alcatraz-a-prison-as-disneyland/
Heizer, J., & Render, B. (2010). *Operations management*. Upper Saddle River: Prentice Hall, Pearson, Inc.
Holden, A. (2008). *Environment and tourism*. New York: Routledge.
Isaac, R. K., & Çakmak, E. (2016). Understanding the motivations and emotions of visitors at Tuol Sleng genocide prison museum (S-21) in Phnom Penh, Cambodia. *International Journal of Tourism Cities*, *2*(3), 232–247.
Jacques, M. M. (2006). *Tracing the holocaust: Experiments in late twentieth-century art and literature (Christian Boltanski, France, Sharon Riis, Canada, Georges Perec, Marguerite Duras, Sarah Kofman)* (Dissertation), University of Alberta, Canada.
Jefferson City Convention and Visitors Bureau. ([2011] 2020). Missouri state penitentiary. *Missouri State Penitentiary History and Timeline*. Retrieved October 12, 2011, from www.missouripentours.com/history/timeline.html (now Retrieved October 31, 2020, from https://www.missouripentours.com/history/)
Jefferson City Convention and Visitors Bureau. (2020). Retrieved February 20, 2020 http://visitjeffersoncity.com/
Kang, E. J., Scott, N., Lee, T. J., & Ballantyne, R. (2012). Benefits of visiting a 'dark tourism' site: The case of the Jeju April 3rd Peace Park, Korea. *Tourism Management*, *33*(2), 257–265.
Kaplan, B. A. (2003). "Aesthetic pollution": The paradox of remembering and forgetting in three holocaust commemorative sites. *Journal of Modern Jewish Studies*, 2(1), 1–18.
Kelley, S. (2017). Beyond neutrality. *American Association for State and Local History (AASLH) History News*, 72(2), 23–27.
Kiely, T. (2011, November 14 and December 12). EVP tourism, San Francisco travel. November 14 and December 12. (D. Bumblauskas, Interviewer).
KOMU News. (2007, June 18). Jefferson City Battle. *KOMU.com*. Retrieved from www.komu.com/news/jefferson-city-battle/
KOMU News. (2010, October 21). Jefferson city unveils plans for conference center. *KOMU.com*. Retrieved February 10, 2020, from www.komu.com/news/jefferson-city-unveils-plans-for-conference-center/
Lennon, J. J. (2017). Dark tourism. *Oxford Research Encyclopedia of Criminology and Criminal Justice*. https://doi.org/10.1093/acrefore/9780190264079.013.212.
Lennon, J. J., & Foley, M. (1999). Interpretation of the unimaginable: The US holocaust memorial museum, Washington, DC, and "dark tourism". *Journal of Travel Research*, *38*(1), 46–50.
Lennon, J. J., & Foley, M. (2000). *Dark tourism: The attraction of death and disaster. Cengage learning EMEA*. London and New York: Continuum.
Missouri Department of Economic Development. (2020). *Missouri State Visitors Bureau*. Missouri State Penitentiary Tours. Retrieved February 10, 2020, from www.visitmo.com/missouri-state-penitentiary-tours.aspx
Missouri State Penitentiary Redevelopment Commission. (2002). *Master/framework plan*. Jefferson City: Parsons HBA and the State of Missouri.

Missouri State Penitentiary Redevelopment Commission. (n.d.). *Missouri state penitentiary redevelopment commission*. Retrieved December 12, 2011, from http://oa.mo.gov/fmdc/dc/msp/index.htbumblausm

News Tribune. (2019a, March 28). *City now accepting Missouri State Penitentiary redeveloper qualifications bids*. Retrieved February 10, 2020, from www.newstribune.com/news/local/story/2019/mar/28/city-now-accepting-missouri-state-penitentiary-redeveloper-qualifications-bids/771828/

News Tribune. (2019b, October 1). Modified Missouri State Penitentiary tours return for fall. Retrieved February 10, 2020, from www.newstribune.com/news/local/story/2019/oct/02/modified-missouri-state-penitentiary-tours-return-fall/797796/

Phelps, N. (2011, December 15). Parks conservancy. (D. Bumblauskas, Interviewer).

Picker, S. (2011, October 31 and November 8). Executive director, Jefferson City convention and visitors bureau. (D. Bumblauskas and J. Fan, Interviewers).

Pirson, M., & Lawrence, P. (2009). Humanism in business—towards a paradigm shift? *Journal of Business Ethics*, *93*(4), 553–565.

Poria, Y., Butler, R., & Airey, D. (2003). Tourism, religion and religiosity: A holy mess. *Current Issues in Tourism*, *6*(4), 40–363.

Public Law 115–391. (2018). *First step act of 2018*. Retrieved February 2, 2020, from www.congress.gov/115/plaws/publ391/PLAW-115publ391.pdf

Randall, A., Ives, B., & Eastman, C. (1974). Bidding games for valuation of aesthetic environmental improvements. *Journal of Environmental Economics and Management*, *1*(2), 132–149.

Reisinger, Y., & Steiner, C. (2006). Reconceptualising interpretation: The role of tour guides in authentic tourism. *Current Issues in Tourism*, *9*(6), 481–498.

Salomone, J. (2004a). A low security prison for women: A best practice in Western Australia. *Forum on Corrections Research*, *16*(1), 34–37.

Salomone, J. (2004b). *Towards best practice in women's corrections: The Western Australian low security prison for women*. Perth: Department of Justice.

Seaton, A. V. (1996). Guided by the dark: From thanatopsis to thanatourism. *International Journal of Heritage Studies*, 2(4), 234–244.

Smith, W. W. (2002). Dark tourism: The attraction of death and disaster-by John Lennon and Malcom Foley. *Annals of Tourism Research*, *4*(29), 1188–1189.

State of Missouri, Division of Design & Construction. (2002). Framework plan for The MSP redevelopment project. Jefferson City: MSP Redevelopment Commission.

State of Missouri Division of Facilities Management. (2006). *MSP redevelopment district design guidelines*. Jefferson City: State of Missouri.

Stone, P. R. (2006). A dark tourism spectrum: Towards a typology of death and macabre related tourist sites, attractions and exhibitions. *Tourism*, *54* (2), 145–160.

Stone, P. R., Hartmann, R., Seaton, A. V., Sharpley, R., & White, L. (Eds.). (2018). *The Palgrave handbook of dark tourism studies* (pp. 335–354). London: Palgrave Macmillan.

Strange, C., & Kempa, M. (2003). Shades of dark tourism: Alcatraz and Robben Island. *Annals of Tourism Research*, *30*(2), 386–405.

Stone, P., & Sharpley, R. (2008). Consuming dark tourism: A thanatological perspective. *Annals of Tourism Research*, *35*(2), 574–595.

Tarlow, P. E. (2007). *Niche tourism: Contemporary issues and trends* (M. Novelli, Ed., pp. 47–59). Burlington: Routledge, Elsevier.

Tinker, E. (2016). *Prisons today: Questions in the age of mass incarceration.*
Tudor, D. T., & Williams, A. T. (2003). Public perception and opinion of visible beach aesthetic pollution: The utilisation of photography. *Journal of Coastal Research*, *19*(4), 1104–1115.
Uhlenbrock, T. (2009, June 3). Missouri prison that once Held Pretty Boy Floyd and James Earl Ray Opens for tours. *Pittsburgh Post Gazette*. Retrieved February 10, 2020, from www.post-gazette.com/pg/09154/969555-37.stm
U.S. Census Bureau and Google Data. (2008, July). *Google public data*. Retrieved December 12, 2011, from www.google.com/publicdata/explore?ds=kf7tgg1uo9ude_&met_y=population&idim=county:06075&dl=en&hl=en&q=san+francisco+ca+population
Visions. (2019, Winter). *Preserving a prison's past*. Ames: Iowa State University Alumni Association.
Wight, A. C., & Lennon, J. J. (2007). Selective interpretation and eclectic human heritage in Lithuania. *Tourism Management*, *28*(2), 519–529.
Yuill, S. M. (2004). *Dark tourism: Understanding visitor motivation at sites of death and disaster* (Master's thesis), Texas A&M University, College Station.

Part III

The Influence of Cultural Context

9 A Grounded Theory Study of Ethnic Tourism in Hainan, China

Shuhui Xing and Dennis Heaton

1. Searching for Humanistic Management in Ethnic Tourism

Humanistic management is concerned with how business promotes human well-being and adds value to society at large. Humanistic management goes beyond the dominant economistic paradigm which assumes that people are materialistic utility maximizers who value individual benefits more than group and societal benefits (Pirson, 2017). A humanistic paradigm shifts concerns about the common good, human flourishing, ethical development, social relationships, and the environment from the margins to the center.

This chapter presents a qualitative exploration of one ethnic tourism project in China which has significantly achieved collaborative relationships with indigenous inhabitants whose culture and native land form the essence of an ethnic-minority theme park. It tells a story of care for the inclusion and economic development of indigenous people while preserving their cultural heritage.

In the sections which follow, we first review some of the literature about ethnic tourism, including prior research studies in China. This is followed by a description of the Bing Lang Gu tourism area and the qualitative research methods used to understand the perceptions of the local ethnic minority people, as well as other stakeholder perceptions, regarding ethnic tourism development. To explore the perceptions of various stakeholders regarding cultural preservation, management of tourism, and balancing social-culture issues in ethnic tourism development, five key stakeholder groups were interviewed: governments, tourism businesses, visitors, ethnic communities, and labor. Grounded theory analysis of the qualitative data led to the construction of an emergent human-centered theory of tourism development. The main analytical finding that is reported is sustainable ethnic tourism development outcomes are achieved when stakeholders share a common philosophy and operate with collaborative relationships. This human-centered model of ethnic tourism development is related to the philosophy of humanistic management and is contrasted to conflict-oriented models from some prior research studies on ethnic tourism in China.

2. Positive and Negative Impacts of Ethnic Tourism

Tourism inherently includes both positive and negative impacts on a community, its economy, and environment (Byrd, 2007). The improper development of tourism may destroy economic, environmental, and social resources (Choi & Sirakaya, 2006; Inskeep, 1991) to exploit present gain while failing to sustain those features of the social and natural environment that are the attraction for tourists.

Nurse (2006) has argued that culture is the fourth dimension of sustainability which is central to the social, economic, and environmental dimensions. For Nurse (2006), the term "culture" includes cultural identities, tangible and intangible heritage, and cultural industries. Protection of culture has been overlooked by hegemonic development approaches which prioritize concepts such as growth, efficiency, and capital accumulation (Nurse, 2006).

Ethnic tourism can enhance awareness of ethnic groups, protect the cultural heritage of local minorities, promote restoration, and protect ethnic attributes that were dying out (Henderson, 2003; MacCannell, 1984). Ethnic tourism has been increasingly supported to attract tourists and to generate income and foreign exchange for ethnic communities (Jamison, 1999; Wood, 1998). Many countries utilize their cultural diversity and use ethnic tourism to stimulate local economic development (Henderson, 2003). Ethnic tourism not only helps to facilitate economic and cultural development but also assists in heritage preservation. Ethnic tourism can enhance awareness of ethnic groups, protect the cultural heritage of local minorities, promote restoration, and protect ethnic attributes that were thought to be dying out (Henderson, 2003; MacCannell, 1984). Moreover, ethnic tourism strengthens the identity of ethnic groups by presenting many opportunities (such as the ethnic culture's arts, performances, and festivals) to display cultures and revive traditions, languages, and cultural pride (Van den Berghe, 1994; Pitchford, 1995; Jamison, 1999).

Although ethnic tourism may help to enhance the economy and bring social benefits to the local community, it also brings negative effects to cultures, their way of life, and their sense of identity (Oakes, 1998; Picard & Wood, 1997; Xie, 2001). A study of indigenous tourism business in Queensland, Australia, concluded that economic rationalism can make tourism enterprises more competitive but may lead to undesirable consequences for indigenous tourism stakeholders (Whitford, Bell & Watkins, 2001). Challenges faced in developing tourism with the indigenous Northern Canadian Cree community, included lack of cultural awareness by tourists, and lack of collaboration among local product suppliers and tourism operators (Notzke, 2004). A case study in Malaysia found that the local Mah Meri community is proud of their culture, but they worry about their natural resources (Kunasekaran, Gill, Talib & Redzuan, 2013).

Ethnicity has become commoditized, which involves selling the re-created ethnic cultural product to tourists. For the tourist market, ethnic tourism commodification can destroy the diverse traditional culture (Oakes, 1993; Smith & Brent, 2001). Because of the high demand for tourism, investors tend to show tourists some "fake" versions of culture, which leads to a loss of authenticity of the local culture (Getz, 1998). This creates an obstacle for sustainable development of the tourism industry. According to Xie (2001), "what the tourist actually sees is just a faint reflection of the true culture" (p. 13). Van den Berghe and Keyes (1984) indicated that tourists can make ethnic minorities less exotic and traditional by transforming an ethnic person into a performer who adjusts his behavior for financial benefit according to what is attractive to tourists.

Van den Berghe and Keyes (1984) used the term "living spectacle" to describe the life of ethnic people who are observed, photographed, and interacted with. Tourists like visiting a "human zoo." Many locals feel that their life is disturbed. Tourists often take photos of them without their permission. Locals are shocked by their unwanted guests, and their children become "spoiled, or develop demeaning begging behavior" (Van den Berghe, 1992, p. 235). The balance between the use of ethnicity as a tourist attraction, the protection of minority cultures, and the promotion of ethnic pride has become an increasing concern (Henderson, 2003; Swain, 1989; Xi, 2001).

3. Ethnic Tourism Studies in China

In the middle and late twentieth century, China's policy shifted to economics over politics, opening China's doors more, and allowing tourism to develop within a socialist market economy model (Zhang, Chong & Ap,1999). As the Chinese economy has improved, travel has been regarded as one of the three "consumption hotspots," along with cars and real estate (Wu, Zhu & Xu, 2000). Consequently, the impact of mass tourism activities in China has grown to exceed the maximum capacity of the natural environment in China (Zhang et al., 1999).

In 2012, the number of domestic tourists reached 2.957 billion, with revenue of $356.567 billion (China National Tourism Administration, 2015). Moreover, this has brought increased damage not only to the ecological environment around tourist attractions (Stronza & Durham, 2008) but also has led to commodification which distorts authentic traditional culture for commercial purposes. Yamamura (2003) found that local minority residents and their culture are changing rapidly as tourism develops in Lijiang, China, a UNESCO World Heritage Site.

Li (2004) investigated the barriers to community tourism in China with a specific case analysis of Nanshan Cultural Tourism Zone (NCTZ) in Hainan Province. Data for the study in NCTZ were collected using observation; reviews of official documents, statistics, printed tourism promotionals, and marketing materials; and in-depth interviews with the provincial and city tourism bureaus, the developer, tourism companies, and community residents. Li explored the perceptions of different stakeholders of community tourism in Hainan and found that the local villagers were excluded from the process of community tourism development. In China, the local government can decide to develop a geographical area by paying compensation to those whose land is to be expropriated (People's Republic of China, 1998). Regarding how tourism development might impact residents' lives, only a little information was available to the local residents. Tosun (2000) has observed that developers may want quick returns from their investment, but community participation would require more time, money, and skills to organize and sustain participation.

Yang and Wall (2009) conducted an empirical study of socio-cultural issues in a well-known tourist destination in Xishuangbanna, Yunnan Province, China. They identified four socio-cultural tensions: authenticity versus cultural commodification, the state regulation versus ethnic autonomy, cultural exoticism versus modernity, and economic development versus cultural preservation. Yang and Wall (2009) explored the relations among four stakeholder groups: governments, ethnic minorities, tourism entrepreneurs, and tourists. The present study expands on that work by including those four stakeholder groups and another group—labor.

Yang and Wall (2009) concluded that the management and planning of ethnic tourism can better achieve balanced development that takes advantage of opportunities while at the same time mitigating impacts. They also mentioned that older minority people were concerned more about cultural changes and advocated for preservation of traditional culture. On the other hand, young minority people were concerned more about making a living, while cultural preservation was less important for them. Many elders believe the market economy has affected their traditional culture, religious beliefs, and minority languages (Yang & Wall, 2009). Yang, Ryan and Zhang (2013) studied the attitudes of Han tourists regarding Tuva minority ethnic tourism in Xinjiang, China. Those researchers observed tensions among stakeholders regarding beliefs, resources, and power.

4. Labor Concerns in Ethnic Tourism

The present research included labor as a stakeholder group in our analysis of stakeholder relations in ethnic tourism. The Global Code of Ethics for Tourism includes Article 9 of Rights of the Workers and Entrepreneurs in the Tourism Industry (World Tourism Organization, n.d.) which mentions that "Partnership and the establishment of balanced relations between enterprises of generating and receiving countries contribute to the sustainable development of tourism and an equitable distribution of the benefits of its growth." China, as a member of the United Nations World Trade Organization (UNWTO), is expected to follow the Global Code of Ethics. Therefore, it is significant to include labor in the stakeholder analysis for ethnic tourism. Moreover, critical management studies (Adler, Forbes & Willmott, 2007; Cunliffe, Forray & Knights, 2002) emphasize understanding business situations from the perspective of workers.

5. Study Area: Bing Lang Gu, Hainan

The majority of the Chinese population is of Han ethnicity. China also has 55 ethnic-minority groups. China is divided into 22 provinces and five autonomous regions: Tibet, Xinjiang, Inner Mongolia, Ningxia, and Guangxi. Autonomous regions are provincial-level administrations, mainly populated by minority people. Minorities make up 8% of the population, totaling about 96 million people and occupying about 65% of China's total area (Sofield & Li, 1998). According to a guideline from the National Tourism Administration in the year 2000, tourism development and cultural policy in China should focus on both heritage and the preservation of ethnic minority cultures (Xie, 2001).

One of the regions in China with a distinctive local culture is Hainan Province, the focus of this research. Hainan is an island in the South China Sea, approximately 25 km off the southwestern coast of the Chinese mainland (Massing, 2016). Hainan's ethnic minorities total 1.4 million people, and Hainan has a wealth of ethnic cultures. The Li minority, with a population of 1.26 million people, is indigenous to Hainan and is the largest of these groups, constituting approximately 15% of the total population of Hainan Province (Xie, 2011). About 60,000 Miao and 7,000 Hui live in Hainan. The ethnic minorities mainly live in the central-south regions of the island and Li and Miao autonomous counties, while the majority Han population is concentrated along the coastal regions (Xie, 2011).

Ethnic minorities generally participate in the informal tourism sector by selling crafts, souvenirs, and local fruit. A small proportion of the population works in the hotels and tourist folk villages. Many folk villagers have been working for the song and dance shows in Hainan. "The most visible manifestation for ethnic employment is in the folk villages where Li work as dance performers and staff" (Xie, 2011, p. 79). Ethnic ceremonies, ethnic foods, and souvenirs provide opportunities for visitors to know more about ethnic cultures.

Bing Lang Gu is an ethnic-minority theme park that exhibits Li and Miao culture. It is about 28 km from Sanya and lies in Baoting Li and Miao Autonomous County, which has the highest number of Li minority communities. Bing Lang Gu covers an area of about 333 hectares and was established in October 1995, during the initial phase of tourism development in Hainan. It is managed as a cooperation between a private businessman and the provincial government. The total number of visitors to Bing Lang Gu was 1.5563 million in 2017, a year-on-year increase of 22.48% (Han & Xi, 2018). The peak of visiting is during Chinese public holidays—Labor Day week, National Day week, and Spring Festival week. The entrance fee is 169 RMB.

Figure 9.1 Li woman doing handicrafts
Source: Authors' own elaboration

At Bing Lang Gu, visitors can see how people lived on the island of Hainan thousands of years ago. Tourists can also watch the process of manufacturing a variety of handicrafts (Figure 9.1), observe how they make fire, watch traditional dances of Li and Miao nationalities, and listen to their songs. The whole of Bing Lang Gu consists of two main parts: the ethnic villages of Li and Miao nationalities and the tropical gardens, where large plantations of the betel nut tree are located. The idea behind Bing Lang Gu was to create a living community that gave tourists the opportunity to learn more about Li culture (Xie, 2011). Moreover, Bing Lang Gu has a rich environment, with rain forests, countless betel nut trees, and wildlife, including lizards, spiders, and monkeys.

6. Methods

The research presented in this chapter is based on the dissertation of the first author. The second author was her principal advisor at Maharishi International University in the United States. The research aimed to understand the perceptions of minority people as well as other stakeholder perceptions regarding ethnic tourism development through an in-depth qualitative study in Hainan island, China. Qualitative research provides an approach that captures the perspectives of various participants with their own expressions, and it allows the researcher to observe the group interactions of an ethnic tourism community.

Interview participants were selected from five stakeholder groups, including government officials, tourists, tourism companies, local Li ethnic-minority members, and the labor of Bing Lang Gu. Five government officials were interviewed first, followed by five employees of tourist companies at Bing Lang Gu. The third and fourth groups of people interviewed for this research were five tourists and five local minority members, respectively, from Bing Lang Gu. Moreover, the fifth group of people interviewed were five workers from Bing Lang Gu. Some participants were recruited through informal conversations in a relaxed atmosphere. For example, in a local minorities store, the researcher began chatting with a local minority woman, and then started to ask her interview questions. After a time, three more Bing Lang Gu employees came in. After talking for a few minutes about their job, the researcher started to ask them interview questions.

The interviews were conducted in Mandarin Chinese for both minorities and other participants because they all know Mandarin. Because the participants were Chinese, the transcript of the interviews was in Chinese. After analyzing the transcripts, English codes were given, and important quotations were put into the final paper in English translation. The number of interviewees was sufficient to reach "saturation," in which the researcher sensed that no new theoretical understanding would be gained by additional data collection.

6.1 Data Collection

The researcher observed the study setting and interviewed participants over two months to build strong relationships with the local community and other stakeholders. According to Creswell and Miller (2000), longer engagement by repeated observation can help researchers build trust with participants so that participants are comfortable sharing more information. Interviews were used to explore in-depth the attitudes and values of different stakeholders.

This research employed "data triangulation," which is the process of comparing data collected from multiple sources, such as field observations, semi-structured interviews, and archival sources (Mathison, 1988). Triangulation can help develop an understanding of potential conceptual categories because different data sources can provide different angles.

The interview questions were constructed based on Browne's (2008) model questions for appreciative inquiry (AI) (Cooperrider, Whitney & Stavros, 2008) to explore deeper personal values about ethnic tourism in Hainan, China. The AI technique has been used in a variety of different areas of tourism (Raymond & Hall, 2008).

After each interviewee read the interview instructions, the researcher asked interview questions based on Browne's (2008) examples of questions in four phases of AI: Discover, Dream, Design, and Destiny. Interviews were semi-structured to be more conversational. An example of a Discover question was: What do you consider some of the most significant trends, events, and developments shaping the future of Bing Lang Gu? An example of a Dream question was: Imagine a time in the future when people look to Bing Lang Gu as an exceptional example of an attractive ethnic tourist destination. Please describe your dream.

6.2 Data Analysis

The study employed grounded theory methods to generate a conceptual model based on a constant comparison method of coding qualitative data (Locke, 2001; Glaser & Strauss, 1967; Corbin & Strauss, 1990). Grounded theory constructs understanding inductively, rather than deductively testing preconceived propositions. Through this inductive method, grounded theory studies may generate insights beyond the prior literature.

Research on the topic of tourism is often discussed by case studies, with emphasis on finding out the descriptions and assertions specific to a place, rather than broader theoretical development. Grounded theory has been a useful method for developing theory in tourism research (Hardy & Beeton, 2005; Daengbuppha, Hemmington & Wilkes, 2006; Kensbock & Jennings, 2011; Humphreys, 2014; Woodside, MacDonald & Burford, 2004; Martin & Woodside, 2008). Grounded theory is particularly useful for studying the effects of tourism on cultures and communities (Hardy & Beeton, 2005). The goal of this research is to create a strong conceptual framework for the improvement of sustainable development of ethnic tourism in minority areas in China. Therefore, grounded theory is a good fit for this research.

In complex social phenomena related to tourism, grounded theory approaches can provide greater insights and understanding (Junek & Killion, 2012). Grounded theory is particularly useful for tourism research and relates to the effects of tourism on cultures and communities (Hardy & Beeton, 2005). Woodside et al. (2004) showed that long interviewing and grounded theory can contribute to understanding the thoughts, decisions, and outcomes from the perspective of leisure travel tourists. Based on such analysis, this study provides a deep understanding of the causes and outcomes of leisure travel. Martin and Woodside (2008) applied long interviews and grounded theory methods to achieve a deep understanding of international foreign visitors' planning processes, motivations, and experiences in Hawaii.

Following the grounded theory procedures presented by Corbin and Strauss (2008), the data were coded in three steps: open coding, axial coding, and selective coding. Coding for this research was recorded using the ATLAS.ti, qualitative data analysis software.

In the first stage of the analysis of data, open coding, the researcher created codes or conceptual labels for events in the data. As additional data were coded, the process of constant comparative analysis compared the similarities and differences between events in the data. In this way, conceptually similar events were grouped together to construct categories and subcategories, which became the elements of the theory. Once identified, categories became the basis for sampling on theoretical grounds. Importantly, during this coding, the researcher constantly wrote memos. As Corbin and Strauss (1990) have explained, open coding uses questioning and constant comparisons and enables the researcher to break through subjectivity and bias.

In the second stage of data analysis, axial coding, the researcher aimed to organize the relationships among the conceptual categories that have been drafted from the first stage. Categories were related through the "coding paradigm" of conditions, context, strategies, and consequences. Questions such as "when, where, why, who, how, and with what consequences" (Massing, 2016) were explored by the researcher. However, the hypothetical relationships proposed during axial coding were considered temporary until the emerging pattern was verified by revisiting the data.

Finally, in the process of selective coding, we looked for a "core" category and categories that need further descriptions with detail. The core category represents the central phenomena of the study. The other categories always closely related to the core category as context, conditions, or consequences. Because discovery of a core category is likely to occur in the later phases of a study, we kept asking questions such as the following from Corbin and Strauss (1990, p. 14):

> What is the main analytic idea presented in this research? If my findings are to be conceptualized in a few sentences, what do I say? What does all the action/interaction seem to be about? How can I explain the variation that I see between and among the categories?

We strove to create a common and abstract core concept to widen the theory's applicability.

7. Results

A total of 158 codes were assigned to different interview transcripts. The codes were then categorized into main themes. Table 9.1 indicates the main categories and their subthemes and the allocation of codes among them.

The following are some of the interview data illustrating selected categories of findings:

Figure 9.2 is the model of ethnic tourism development which was constructed through the axial coding and selective coding processes of grounded theory analysis. The core category, in the center of Figure 9.2,

Table 9.1 Codes within the main categories

Ethnic tourism development			
Cultural protection	*Future trend*	*Tourism product*	*Government support*
Cultural heritage Organization	Combine with fashion show	DIY silverware	Control planning of development
Display the minority's culture	Build an education base	Homestay with minority families	Cares about local minority
Explore culture	Cooperate with rural tourism	Miao medicine	Train government officials
Learn and pass on culture and skills	Establish minority middle school	Wedding celebration	Fund support
		Customized service	Support local to pass on culture
Teach tourist minority's culture	Focus on experiential tourism	Well-designed ethnic show	Provide more favorable policy
Establish protection awareness	Combine farming culture	Outward development	
Keep the minority old building	Improve the minority's show	Green food	
Protect elderly Li women	Looking for investment		
Cultural values	Let the guest feel at home	***Minority awareness***	
Management treats employees well	Use technology to display culture	Minority work in management	Minority autonomy
		Enhance self-identity and pride	Minority can build a positive image for world
Help one another	Cooperate with international and tourism companies	Minority should Love their culture	
Care about local minority			Minority display their cultures internationally
Local minority support tourism		Encourage minority to participate more	
Relationship with minority (win-win mode)		***Core ideology***	***Talent recruitment***
Good for minority	***Good for Bing Lang Gu***	Authenticity	Combine local and non-local talents
Minority has five different incomes	Effective management	Creativity	Hire more minority employees
	Environmentally friendly	Human-based management	Hire cultural protection expert
Visit poor families	Local minority nice to tourists	Sustainable development	Training employees
Year-end bonus	Suggestion boxes are available		Hire talents with strong experience
Improve life standard of minority	Well-designed show structure	***Vision***	***Culture promotion***
Minority can keep their plant	Minority support the management	Internationalization	Combine traditional and modern activities
Provide minority free stores		Create a brand	
Provide work opportunities		Diversification	Create cultural environment
Support children to go to college		Integration of resources	Enhance advisement promotion
Year-end dinner celebration			Expand market and more events
			More cultural products
			Improve service quality

Source: Authors' own elaboration

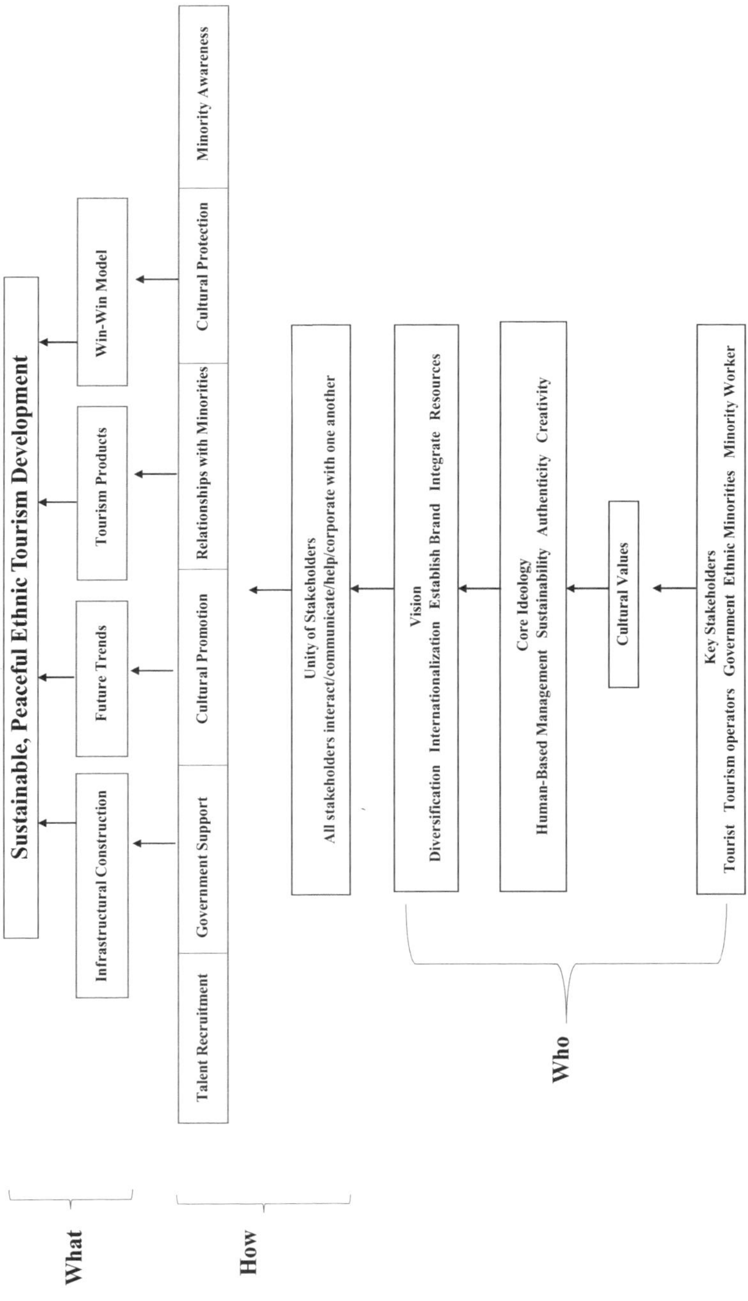

Figure 9.2 Model of ethnic tourism development derived from grounded theory analysis

is unity of stakeholders—the finding in this case that all stakeholders communicated and cooperated with one another. The analysis of this research study in Bing Lang Gu suggests that this is the key factor for cooperatively realizing sustainable, peaceful ethnic tourism development.

As depicted in this theoretical model, cultural values play a key role in bringing about a shared ideology and vision among the stakeholder groups. Based on the cultural values of stakeholders, which start from the bottom, stakeholders share the same core ideology (human-based management, sustainability, authenticity, creativity). Having the same core ideology further shaped the common vision (diversification, internationalization, establish brand, integrate resources) among different stakeholders.

Furthermore, because the stakeholders share a common ideology and vision, the key stakeholders (Tourist, Tourism Operator, Government, Ethnic Minorities, Minority Worker) are able to communicate and cooperate with one another. Cultural harmony in stakeholder interactions is enabled by having a shared vision. Otherwise, they would not be helping one another. Because stakeholders have the same vision, they can work together.

Based on the unity of stakeholders, stakeholders have practical plans and actions—Talent Recruitment, Government Support, Cultural Promotion, Relationships with Minorities, Cultural Protection, and Minority Awareness. This practical planning leads to more detailed actions such as Government Support leading to Infrastructural Construction, Cultural Promotion leading to Future Trends, Relationships with Minorities leading to Tourism Products, and Cultural Protection leading to a Win-Win Model. In the end, all these practical plans and actions lead to what we labeled as sustainable, peaceful ethnic tourism development.

8. Stakeholder Perspectives

Interview findings from the various stakeholder groups provided ideas that can contribute to the furtherance of sustainable ethnic tourism development for Bing Lang Gu. Tourists suggested that Bing Lang Gu should create more activities for them and increase interaction with them. Moreover, tourists suggested that Bing Lang Gu should improve the authenticity of their culture. Tourists want more experiential programs and more cultural products.

Tourism operators in Bing Lang Gu want to meet the needs of tourists by providing more experiential tourism products and increasing interactions with tourists. In addition, tourism operators aim to continue working on protection of the traditional culture and relationship with the indigenous people. Tourism operators' wish for the local ethnic population is that they can love their cultures by enhancing their self-identity. The vice president of Bing Lang Gu pointed out: "If they wish to inherit their culture, then they need to start from knowing themselves." The HR manager of Bing Lang Gu mentioned: "Firstly, we should organize the local students to study the handicraft skills, which will let more people know about this special skill because this skill right now is only known by some old Li women; it's dying out."

Government representatives aim to preserve the minority culture and improve the living standard of ethnic minorities. Moreover, the government encouraged tourists and employees to provide more suggestions for planning. The government also suggested that the indigenous people should change to experiential tourism and hire more talents. One government official said: "Local minorities lack awareness to give suggestions and lack awareness to join the tourism planning. Maybe they did not realize their right. Mainly the tourism companies and government play a leadership role in planning." Another government official mentioned: "Bing Lang Gu has an area to sell the minority handmade silverware to tourist, however, Bing Lang Gu management should let tourist know how they make silverware, let them create their own silverware."

Findings regarding the perspectives of local minority people include wishing to improve their living standard. Further, local minority people are nice to tourists and support the management of tourism operators. Moreover, the local minority agree on cultural protection. However, some minority people lack self-identity and are not familiar with their culture and tradition.

Minority workers appreciate what management did for them and the whole community. They follow the core ideology of the management team and share the same visions. Minority workers wish to protect elderly Li women because they think these women are the best precious gift for Bing Lang Gu. Bing Lang Gu provides many work opportunities for the local people. A local Li woman said, "As soon as you want to come, Bing Lang Gu will provide a job for you." The vice president of Bing Lang Gu pointed out:

> Now we are increasing the number of local minority employees, from age of 18 to age of 80, different age group of local people will arrange in different positions, young people can be a tourism guide or security, old people can make handicraft, you can do whatever you good at, even if you only good at farming. . . . We are combining farming culture with our service.

9. Implications for Practice

From the interviews and observations of our research at Bing Lang Gu, here are some recommendations for the site, which may be applicable for other sites as well. Because sustainable development of the ethnic tourism site depends on continued collaboration of five stakeholders—tourists, tourism operators, minority workers, government, indigenous community—there are recommendations for each of these stakeholder groups.

For tourists: Tourists are encouraged to provide their suggestions to tourism operators by submitting their suggestions to the "suggestion box" at the sites. The social media of tourists can help to promote the image of Bing Lang Gu. Feedback of their experiences is also important for tourism operators to improve their services.

For tourism operators: Tourists seem interested in experiencing the traditional activities of the local community. Indigenous cultural events, crafts workshops, indigenous dance shows, and instruction in minority languages can encourage the tourist to interact with the indigenous community and engage more with indigenous activities.

Current research found that tourism product ideas can come from tourists and suppliers of tourism products. Structure communication between the tourists and tourism operators can enhance product development.

Ethnic tourism sites should cooperate with more international companies; in this way, their cultures can be known by more countries. Moreover, tourism operators should use advanced technology to display ethnic culture (such as advanced museum display video equipment and virtual reality [VR] experiences).

The study shows that tourism operators support the indigenous community by hiring minority people and providing scholarships to their children. This financial support maintains a good relationship with the indigenous community.

For local minority workers: Cultural values training courses for local ethnic employees can keep their cultural values strong and enhance appreciation of their culture's values. Providing management and leadership training for employees is essential to improve their knowledge and leadership.

For the indigenous community: Self-identity can motivate the indigenous community to maintain traditional events and showcase their cultures to tourists. The traditional handicraft skills of elderly Li minority women need to be passed on to maintain the cultural attractiveness of the site. The local community should encourage their children and students to learn their traditional skills, especially the worldwide intangible cultural heritage handicraft skills, so the cultures can pass on.

Many stores are owned by local minority people at Bing Lang Gu. They should ensure their product's quality meets the standards and displays the traditional creative design of the indigenous product.

For government: Because Bing Lang Gu displays ten of the worldwide intangible cultural heritage skills, the government has supported the site in expanding tourism development through funds support and beneficial policies. In the future, the government should continue to support tourism development by providing more beneficial promoting policies and protection of the indigenous community.

These recommendations are in line with those reported by Nurse (2006) from the International Meeting to Review the Implementation of the Programme of Action for the Sustainable Development of Small Island Developing States.

10. Conclusions

The case of Bing Lang Gu illustrates a humanistic approach (Pirson, 2017) to tourism in that support for indigenous people and preservation of authentic culture are genuinely valued. Decisions are based on more than a purely economic perspective; culture and the natural environment are not merely "resources" to be turned into profit but are held to be valued ends in their own right. This case does illustrate "moving towards more significant attention to the needs of human subjects, both tourists, and inhabitants of the tourist places," which the editors of this volume have called for (Giudici & Della Lucia, 2018).

Maintaining dialog with all stakeholders is one of the core principles of humanistic management (Melé, 2016). The central concept emerging from our grounded theory analysis was Unity of Stakeholders. This finding contrasts with the research by Yang and Wall (2009), that highlighted tensions around social-cultural issues in Chinese ethnic tourism development. Current research found that, in the case of Bing Lang Gu, a strong cultural values environment enabled stakeholder groups to share a common vision and ideology. As a result, stakeholder groups can better interact with, communicate with, and help one another, which leads to the planning and execution of ethnic tourism products and the overall outcome of sustainable ethnic tourism development.

References

Adler, P. S., Forbes, L. C., & Willmott, H. (2007). Critical management studies. *The Academy of Management Annals*, *1*(1), 119–179.

Aziz, R. C., Abdul, M., Aziz, Y. A., & Rahman, A. A. (2013). Appreciative Inquiry: An alternative research approach for sustainable rural tourism development. *Journal of Tourism, Hospitality & Culinary Arts*, *5*(2), 1–18.

Browne, B. W. (2008). *Crafting appreciative questions. A how to guide*. Chicago: Imagine Chicago. Retrieved from www.imaginechicago.org/docs/ai/Crafting%20Appreciative%20Questions.doc

Byrd, E. T. (2007). Stakeholders in sustainable tourism development and their roles: Applying stakeholder theory to sustainable tourism development. *Tourism Review*, *62*(2), 6–13.

China National Tourism Administration. (2015). *China domestic tourism sample survey*. Beijing, PRC China: CNTA.

Choi, H. C., & Sirakaya, E. (2006). Sustainability indicators for managing community tourism. *Tourism Management*, *27*(6), 1274–1289.

Cooperrider, D. L., Whitney, D. K., & Stavros, J. M. (2008). *Appreciative Inquiry handbook: For leaders of change*. Brunswick, OH: Berrett-Koehler Publishers.

Corbin, J. M., & Strauss, A. (1990). Grounded theory research: Procedures, canons, and evaluative criteria. *Qualitative Sociology*, *13*(1), 3–21.

Corbin, J. M., & Strauss, A. (2008). *Basics of qualitative research: Techniques and procedures for developing grounded theory* (3rd ed.). Thousand Oaks, CA: Sage.

Creswell, J. W., & Miller, D. L. (2000). Determining validity in qualitative inquiry. *Theory into Practice*, *39*(3), 124–130.

Cunliffe, A., Forray, J. M., & Knights, D. (2002). Considering management education: Insights from critical management studies. *Journal of Management Education*, *26*(5), 489–495.

Daengbuppha, J., Hemmington, N., & Wilkes, K. (2006). Using grounded theory to model visitor experiences at heritage sites: Methodological and practical issues. *Qualitative Market Research: An International Journal*, *9*(4), 367–388.

Dierksmeier, C. (2016). What is 'humanistic' about humanistic management? *Humanistic Management Journal*, *1*(1), 9–32.

Getz, D. (1998). Event tourism and the authenticity dilemma. In W. F. Theobald (Ed.), *Global tourism* (pp. 409–427). Oxford: Butterworth—Heinemann.

Giudici, E., & Della Lucia, M. (2018). *Book proposal: Shaping a humanistic perspective for the tourism industry*. Retrieved from www.societamanagement.it/wp-content/uploads/book-proposal-shaping-a-humanistic-perspective-for-the-tourism-industry.pdf

Glaser, B., & Strauss, A. (1967). *The discovery of grounded theory*. Chicago, IL: Aldine.

Han, T., & Xi, X. (Eds.). (2018, January 2). *The number of tourists in Hainan Scenic Spot increased in 2017*. Retrieved February 15, 2020, from http://hi.people.com.cn/n2/2018/0102/c231190-31097676.html

Hardy, A., & Beeton, R. J. S. (2005). Using Grounded Theory to explore stakeholder perceptions of tourism. *Tourism and Cultural Change*, *3*(2), 1–24.

Hayward, P., & Li, J. F. (2010). Gilding the pearl: Cultural heritage, sexual allure and polychromatic exoticism on Hainan island. *Perfect Beat*, *11*(2), 119–140.

Henderson, J. (2003). Ethnic heritage as a tourist attraction: The Peranakans of Singapore. *International Journal of Heritage Studies*, *9*(1), 27–44.

Hitchcock, R. K., & Brandenburgh, R. L. (1990). Tourism, conservation, and culture in the Kalahari Desert, Botswana. *Cultural Survival Quarterly*, *14*(2), 20–24.

Humphreys, C. (2014). Understanding how sporting characteristics and behaviours influence destination selection: A grounded theory study of golf tourism. *Journal of Sport & Tourism*, *19*(1), 29–54.

IGI Global. (n.d.). *What is substantive theory*. Retrieved August 4, 2017, from www.igi-global.com/dictionary/substantive-theory/28672

Inskeep, E. (1991). *Tourism planning: An integrated and sustainable development approach*. New York: Van Nostrand Reinhold.

Jamison, D. (1999). Tourism and ethnicity: The brotherhood of coconuts. *Annals of Tourism Research*, *26*(4), 944–967.

Johnson, S. D. (1995, Spring). Will our research hold up under scrutiny? *Journal of Industrial Teacher Education*, *32*(3), 3–6.

Johnston, L. (2001). (Other) bodies and tourism studies. *Annals of Tourism Research*, *28*(1), 180–201.

Junek, O., & Killion, L. (2012). Grounded theory. In L. Dwyer, A. Gill, & N. Seetaram (Eds.), *Handbook of research methods in tourism: Quantitative and qualitative approaches* (pp. 325–338). Northampton, MA: Edward Elgar.

Kensbock, S., & Jennings, G. (2011). Pursuing: A grounded theory of tourism entrepreneurs' understanding and praxis of sustainable tourism. *Asia Pacific Journal of Tourism Research, 16*(5), 489–504.

Klieger, P. C. (1990). Close encounters: "Intimate" tourism in Tibet. *Cultural Survival Quarterly, 14*(2), 38–42.

Kunasekaran, P., Gill, S. S., Talib, A. T., & Redzuan, M. R. (2013). Culture as an indigenous tourism product of Mah Meri community in Malaysia. *Life Science Journal, 10*(3), 1600–1604.

Li, Y. (2000). Geographical consciousness and tourism experience. *Annals of Tourism Research, 23*(4), 863–883.

Li, Y. (2004). Exploring community tourism in China: The case of Nanshan cultural tourism zone. *Journal of Sustainable Tourism, 12*(3), 175–193.

Locke, K. (2001). *Grounded theory in management research*. Chicago, IL: Sage.

MacCannell, D. (1984). Reconstructed ethnicity tourism and cultural identity in third world communities. *Annals of Tourism Research, 11*(3), 375–391.

Martin, D., & Woodside, A. G. (2008). Grounded theory of international tourism behavior. *Journal of Travel & Tourism Marketing, 24*(4), 245–258.

Massing, K. (2016). *Finding an ecomuseum ideal for Hainan Province: Encouraging community participation in intangible cultural and natural heritage protection in a rural setting in China* (Doctoral dissertation), Newcastle University. Retrieved from https://theses.ncl.ac.uk/dspace/bitstream/10443/3344/1/Massing,%20K.%202016.pdf

Mathison, S. (1988). Why triangulate? *Educational Researcher, 17*(2), 13–17.

Melé, D. (2016). Understanding humanistic management. *Humanistic Management Journal, 1*(1), 33–55.

Muller, H. (1994). The thorny path to sustainable tourism development. *Journal of Sustainable Tourism, 2*(3), 131–136.

Notzke, C. (2004). Indigenous tourism development in southern Alberta, Canada: Tentative engagement. *Journal of Sustainable Tourism, 12*(1), 29–54.

Nurse, K. (2006). Culture as the fourth pillar of sustainable development. *Small States: Economic Review and Basic Statistics, 11*, 28–40.

Oakes, T. S. (1993). The cultural space of modernity: Ethnic tourism and place identity in China. *Environment and Planning D: Society and Space, 11*(1), 47–66.

Oakes, T. S. (1998). *Tourism and modernity in China*. London: Routledge.

People's Republic of China. (1998). *Law of the people's republic of china on land management*. Promulgated 29 August 1998 as order No. 8 of the president of the People's Republic of China. Beijing: National People's Congress.

Picard, M., & Wood, R. E. (1997). *Tourism, ethnicity, and the state in Asian and pacific societies*. Hawaii: University of Hawaii Press.

Pirson, M. (2016). Welcome to the humanistic management journal. *Humanistic Management Journal, 1*(1), 1–7.

Pirson, M. (2017). A Humanistic Perspective for Management Research: Protecting dignity and promoting well being. *Humanistic Management Association, Research Paper Series*, 17–18.

Pitchford, S. (1995). Ethnic tourism and nationalism in Wales. *Annals of Tourism Research*, *22*(1), 35–52.

Punch, K. (2005). *Introduction to social research: Quantitative and qualitative approaches*. London: Sage.

Schein, E. H. (1984). Coming to a new awareness of organizational culture. *Sloan Management Review, 25*(2), 3–16.

Schein, E. H. (2006). *Organizational culture and leadership* (Vol. 356). John Wiley & Sons.

Smith, V. L., & Brent, M. (2001). *Hosts and guests revisited: Tourism issues of the 21st century*. New York: Cognizant Communication Corp.

Sofield, T. H., & Li, F. M. S. (1998). Tourism development and cultural policies in China. *Annals of Tourism Research*, *25*(2), 362–392.

Sorensen, A. (2003). Backpacker ethnography. *Annals of Tourism Research*, *30*(4), 847–867.

Spencer-Oatey, H., & Franklin, P. (2012). What is culture. *A Compilation of Quotations. GlobalPAD Core Concepts*.

Strauss, A., & Corbin, J. (1998). *Basics of qualitative research techniques*. London: Sage.

Stronza, A., & Durham, W. H. (Eds.). (2008). *Ecotourism and conservation in the Americas* (Vol. 7). Wallingford, Oxfordshire: CABI.

Swain, M. B. (1989). Developing ethnic tourism in Yunnan, China: Shilin Sani. *Tourism Recreation Research*, *14*(1), 33–39.

Tan, W., & Wang, P. (2013). Appreciative Inquiry in Chinese cultures: Philosophy and practice. *AI Practitioner*, *15*(3), 31–39.

Taylor, J. P. (2001). Authenticity and sincerity in tourism. *Annals of tourism research*, *28*(1), 7–26.

Thompson, R. (2003). *Cultural aspects of success in strategic, international distance education collaborations in the English-speaking Caribbean* (Unpublished doctoral dissertation), Maharishi University of Management, Fairfield, Iowa.

Tosun, C. (2000). Limits to community participation in the tourism development process in developing countries. *Tourism management*, *21*(6), 613–633.

Trousdale, W. J. (1999). Governance in context: Boracay Island, Philippines. *Annals of Tourism Research*, *26*(4), 840–867.

Van den Berghe, P. (1992). Tourism and the ethnic division of labor. *Annals of Tourism Research*, *19*(2), 234–249.

Van den Berghe, P., & Keyes, C. (1984). Introduction: Tourism and re-created ethnicity. *Annals of Tourism Research*, *11*(3), 343–352.

Whitford, M., Bell, B., & Watkins, M. (2001). Indigenous tourism policy in Australia: 25 years of rhetoric and economic rationalism. *Current Issues in Tourism*, *4*(2–4), 151–181.

Wood, R. (1998). Touristic ethnicity: A brief itinerary. *Ethnic and Racial Studies*, *21*(2), 218–241.

Woodside, A. G., MacDonald, R., & Burford, M. (2004). Grounded theory of leisure travel. *Journal of Travel and Tourism Marketing*, *17*(1), 7–39.

World Tourism Organization. (n.d). *Global code of ethics for tourism-Article 9*. Retrieved May 8, 2017, from http://ethics.unwto.org/en/content/global-code-ethics-tourism-article-9

Wu, B., Zhu, H., & Xu, X. (2000). Trends in China's domestic tourism development at the turn of the century. *International Journal of Contemporary Hospitality Management*, *12*(5), 296–299.

Xie, P. F. (2001). *Authenticating cultural tourism: Folk villages in Hainan, China* (doctoral dissertation), University pf Waterloo. Retrieved from https://uwspace.uwaterloo.ca/bitstream/handle/10012/975/fxie2001.pdf?sequence=1&isAllowed=y

Xie, P. F. (2011). *Authenticating ethnic tourism* (Vol. 26). Bristol, Buffalo: Toronto: Channel View Publications.

Yamamura, T. (2003). Indigenous society and immigrants: Tourism and retailing in Lijiang, China, a world Heritage city. *TOURISM: An International Interdisciplinary Journal*, *51*(2), 215–235.

Yang, J., Ryan, C., & Zhang, L. (2013). Social conflict in communities impacted by tourism. *Tourism Management*, *35*, 82–93.

Yang, L., & Wall, G. (2009). Ethnic tourism: A framework and an application. *Tourism Management*, *30*(4), 559–570.

Zhang, Q. H., Chong, K., & Ap, J. (1999). An analysis of tourism policy development in modern China. *Tourism Management*, *20*, 471–485.

10 Positive Impacts of Sustainable Tourism

Preservation of Historical Languages and Practices

Serena Lonardi, Umberto Martini, and John S. Hull

1. Introduction

Sustainable tourism has taken a considerable role in the literature, ever since the UN Conference on Environment and Development of Rio de Janeiro was held in 1992 and the 'Agenda 21 for the Travel and Tourism Industry' was discussed and approved in 1997. It is a widely accepted fact nowadays that, when responsibly managed, tourism can positively affect not only the economy of a country but also its society and culture (Archer, Cooper & Ruhanen, 2005). Sustainable activities, including tourism, aim at preserving natural and cultural environments for future generations (UNESCO, 2003; Soini & Birkeland, 2014). This is even more urgent for practices that will disappear if not practiced daily, such as intangible heritage in general and endangered languages in particular (Kim, Whitford & Arcodia, 2019; Maffi, 2007).

This chapter analyzes the role of tourism in preserving and revitalizing the cultural identity of minority groups and especially their traditional languages. Previous research has proved that minority languages, being the representation of a unique culture (Krauss, 1992), are an asset and a differentiating factor for the destination (Kelly-Holmes & Pietikäinen, 2014; Whitney-Squire, 2016). Tourists' genuine interest in the cultural heritage of a group then promotes a sense of pride in their cultural background, triggering a sense of identity and a desire to learn more, thus creating a virtuous circle of culture and language preservation (Greathouse-Amador, 2005; Whitney-Squire, 2016).

This chapter analyzes whether and under which circumstances tourism can contribute to the process of revitalization of endangered minority languages and traditional practices. To achieve a better understanding of the phenomenon studied, three case studies are compared from two Indigenous groups from British Columbia, Canada (Little Shuswap Lake Indian Band and the Okanagan Indian Band) and one from the Cimbrian people in Giazza, Italy. All cases are tourism destinations, where residents are members of a minority community and speak partially in an endangered language. The considerable differences between the cases, explained more in detail in a later section of this chapter, allow for a significant and fruitful comparison.

The chapter is structured as follows: the following section summarizes the literature review and presents the conceptual framework. Then, the methodology is explained, together with the selected case studies. Finally, the results are presented and discussed.

2. Literature Review and Conceptual Framework

2.1 Minority Languages and Tourism

According to UNESCO (2003), a language becomes endangered when fewer and fewer people speak it and do not pass it on to future generations. UNESCO's *Atlas of the World's Languages in Danger* (Moseley, 2010) identified 2,464 endangered languages out of 6,000 spoken today in the world. Various factors have been identified as causes of language loss. Generally, people stop speaking a language when another language is seen as more prestigious, even though it is not prestige alone that brings about a language shift. The benefits of speaking a language are often due to the economic factors associated with their use (Mufwene, 2002). Not surprisingly, this often results in minority languages—that is, those languages spoken by a minority of a population—being threatened (Capotorti, 1977). Because each language stands for a unique worldview and is the vehicle for collective memories and values, losing it could then lead to terrible consequences for a people and their cultural identity (Fishman, 1991).

It follows that language revitalization is important and urgent (Fishman, 1991). It can obviously be achieved through different methods and theories. Of course, linguists need to take a primary role in this (Fishman, 1991; Krauss, 1992). However, tourism can be seen as a tool to preserve and revitalize endangered languages. Previous research focused on endangered Indigenous languages in North America argues that tourism can be an efficient tool to preserve and revitalize minority languages, because tourists now look for authentic and meaningful experiences regarding the local culture of the place they are visiting (Kim, Ritchie & McCormick, 2012; Whitney-Squire, 2016). They see their holiday as an opportunity to learn something and experience a different culture and lifestyle, and therefore the cultural heritage of a population, including its traditional language, can become a pull factor for tourism (Ritchie, Cooper & Carr, 2003; Ugolini & Costa, 2009).

Considering this framework, the role played by the tourism planning processes is undoubtedly relevant, as regards, in particular, the capacity to build a common strategy among economic actors, institutions, and residents, with the aim to add value to local tourist offerings through the preservation, use and "commodification" of the local language (see

Heller, 2003). Tourist offerings must be created and managed in a way that allows local language—intended as one of the main vehicles for the collective memory and values of a population—to be turned into authentic tourism experiences, while protecting intrinsic cultural value and cultural identity (Kelly-Holmes & Pietikäinen, 2014; Whitney-Squire, 2016).

This means that destination managers and local institutions must develop a strategy focused on authenticity and experientiality, involving different categories of local actors and businesses in a common project of territorial value creation, in a sort of "cultural heritage marketing" process (Kolar & Zabkar, 2010; Qilou, Zhang, Zhang & Ma, 2015). This authenticity must be perceived by guests as an innovative and differentiating component of the tourist products of the territory and not as a form of "living pasts" or as a sort of virtual representation of a virtual (but false) reality (see Zhu, 2012; Knudsen, Rickly & Vidon, 2016).

The introduction of traditional languages and culture in tourism has raised concerns and doubts related to the authenticity of the cultures represented in tourism (MacCannell, 1973; Duchêne & Heller, 2012). Cohen (1988), however, argued that commodification may actually contribute to maintenance and revitalization of the local culture and identity by generating interest and value, not only for tourists but also for Indigenous communities themselves. If responsibly and sustainably managed, it thus may help to preserve traditions, which would otherwise slowly disappear, including the traditional languages. This means the community needs to be involved in the decision-making process and in the creation of contents (Whitney-Squire, 2016).

Previous studies that analyzed the relationship between minority languages and tourism considered languages as artifacts displayed in museums, like the National Museum of the Finnish Sámi (Kelly-Holmes & Pietikäinen, 2016) and the Afrikaans Language Museum (Burden 2007), and the labeling of souvenirs (Pietikäinen, Kelly-Holmes & Peer, 2011). Other research analyzed the linguistic landscape, that is, visibility of the language in public spaces, and selected the local airport (Heinrich, 2010), brochures, and social media (Whitney-Squire, 2016) or signs indicating touristic routes (Moriarty, 2014). Finally, some authors looked at the linguistic soundscape, that is, the use of the language in tours and to greet guests (Kelly-Holmes & Pietikäinen, 2014; Whitney-Squire, 2016).

The great interest of tourists in knowing more about traditional languages eventually promoted a sense of pride in being Indigenous, thus motivating preservation and revitalization of the language (Greathouse-Amador, 2005; Whitney-Squire, 2016). For instance, in Haida Gwaii (British Columbia), tourism planners developed a series of projects that included the Haida language, such as paying staff to learn the language; encouraging tour guides to use it, such as in greeting visitors; and working with elders to develop tour content. There were evening language talks with visitors, and efforts were made to increase language content in interpretive and promotional materials and in social media. This approach to integrate traditional language was very successful since tourists really appreciated this unique value. Moreover, the use of the traditional language in tourism led to increasing awareness of people's origins, making them proud of that, so that, in turn, they were encouraged to learn more (Whitney-Squire, 2016).

2.2 Theoretical Framework

The literature review summarized here shows how language and tourism experiences are intrinsically interrelated. On one hand, minority languages contribute to the attractiveness of the destination and, therefore, represent a competitive advantage (Kelly-Holmes & Pietikäinen, 2014; Moriarty, 2014; Pietikäinen et al. 2011; Whitney-Squire, 2016). On the other hand, including those languages in tourism experiences often leads to enhanced pride among community members and, therefore, to the desire to learn more about their cultural heritage and traditional language (Greathouse-Amador, 2005; Whitney-Squire, 2016). Tourism creates a virtuous circle of language and cultural preservation, thus reinforcing the distinctive value of the destination.

3. Case Description and Methodology

The following paragraphs present the case studies analyzed: Indigenous people in British Columbia, with a focus on the Little Shuswap Lake Indian Band and the Okanagan Indian Band, and the Cimbrian people in Giazza, Italy.

The authors were first involved in an empirical research project at Thompson Rivers University (Kamloops, British Columbia, Canada), where they considered the two Indigenous communities that are briefly presented in the following paragraphs. Because some significant results were made, they decided to consider another case in Italy, to analyze whether completely different socio-cultural backgrounds would produce similar outcomes concerning the relationship between tourism and language and culture preservation.

3.1 Indigenous People in British Columbia

The most accepted definition of Indigenous people is the one approved by The United Nations Working Group on Indigenous Populations (Cobo, 1986):

> Those [people] which, having a historical continuity with pre-invasion and pre-colonial societies that developed on their territories,

> consider themselves distinct from other sectors of the societies now prevailing on those territories, or parts of them.
>
> (p. 10)

In Canada, Indigenous peoples consist of First Nations, Inuit, and Métis and comprise 2.84% of the population, or approximately 1.6 million people (Statistics Canada, 2016). British Columbia is Canada's westernmost province, where Indigenous people represent 5.93% of the population (Statistics Canada, 2016). European settlers first arrived at the end of the 18th century, resulting in Indigenous peoples having to face massive changes that would significantly transform their lifestyles. To begin, diseases spread for the first time among Indigenous communities, devastating the populations and thus contributing to weakening and undermining their culture (Mason, 2014). But they also had to face other massive lifestyle transformations. Most importantly, their lands were confiscated, and Indigenous people were forced to permanently live on reserves, where they had to work hard and were not allowed to speak their language or engage in their traditional practices, while the government pursued policies of assimilation and repression (Mason, 2014).

Another example of assimilation can be found in the education system. When the education of Indigenous children came under colonial control, they were abruptly removed from their families and obliged to live in residential schools (Young, 2015). There, children experienced terrible emotional, physical, and psychological abuse. They were taught that their rituals were evil, their relatives savage, and their languages primitive, which made them feel ashamed of their own cultures and origins. The children were obliged to speak only French or English, and their native languages were outlawed. As a consequence, they rejected their culture, and many even forgot their mother tongues, which eventually led to cultural genocide and an almost irreparable language loss (Partridge, 2011). In fact, in most of the cases, even people who could remember their ancestral language would not pass it on to their children, to prevent them from having the same terrible experience (Crawford, 1995).

Stereotypes of American Indian people have been an important part of the cultural imaginary in Europe since Columbus discovered the New World (Fleming, 2006). Indigenous peoples have been described both as "noble savage" (Stirrup, 2013, p. 6) and uncivilized cannibals. In the late 19th and early 20th centuries, Western people learned about American Indian people through the literature, through traveling performances, and later on, through the cinema (Stirrup, 2013). Not surprisingly, the fascination of Europeans for the Indigenous peoples of North America, together with the fact that tourists have always been looking for something completely different from their everyday experiences, has naturally led to the origins of Indigenous tourism (Butler & Hinch, 2007). Indigenous tourism has been defined as "all tourism businesses majority owned, operated and/or controlled by [Indigenous] peoples that can demonstrate a connection and responsibility to the local Aboriginal community and traditional territory where the operation resides" (Aboriginal Tourism Association of Canada, 2016, p. 4).

British Columbia is well known for its stunning and diverse natural environment, which allows visitors to practice a wide variety of outdoor sports, such as hiking, canoeing, skiing, and snowshoeing. However, British Columbia is becoming more and more popular also for its millennial Indigenous cultural heritage (Province of British Columbia, 2011).

In 2016–2017, more than 400 companies related to Indigenous tourism were operating in British Columbia, mostly working in retail (34%), outdoor (19%), and accommodation (12%) (Indigenous Tourism BC, 2018). They generated approximately $705 million in gross domestic output and created 7,400 jobs. The greatest portion of travelers participating in Indigenous tourism experiences are domestic, that is, Canadian residents, whereas the remaining visitors come from China, Germany, the United Kingdom, and the United States (Indigenous Tourism BC, 2018).

A single and striking example worth mentioning here is the Vancouver 2010 Winter Olympics, where Indigenous groups were involved in the development and organization of the event. In particular, the four First Nations groups on whose traditional lands the Olympics were hosted were recognized as official partners and played an important role in the opening ceremony. Even though some Indigenous people took a position against the Olympics, for others, the Olympics were regarded as a perfect opportunity to show their culture to the world and have an important role in key decisions (BBC news, 2010). In fact, the 2010 Olympic and Paralympic Winter Games raised both national and international interest for Indigenous tourism experiences. As a result, customer demand has been growing since then (The Aboriginal Tourism Association of British Columbia, 2015).

The two Indigenous groups of British Columbia considered in this study are the Little Shuswap Lake Indian Band and the Okanagan Indian Band. The Little Shuswap Lake Indian Band is a First Nations band located in the central interior of British Columbia. Their main reserve is in Chase, near the Little Shuswap Lake. Their traditional language is the Secwepemctsín language, which is categorized as severely endangered by UNESCO's *Atlas of the World's Languages in Danger* (Moseley, 2010), that is, estimated speakers left number approximately 1,100 and on average they are 50 years old. In 1992, the Little Shuswap Lake Indian Band opened the Quaaout Lodge, a tourism resort situated on the shores of the lake. Since that time, other companies were founded by the band aimed at improving the local economy, as well as on developing tourism in the town, like the conference center (2001), Talking Rock Golf (2007), Le7ke Day Spa (2011), Little Shuswap Lake Gas and Lakehead Helicopters (www.lslib.com).

The Okanagan Indian Band is located in the southern interior of British Columbia and north-central Washington (US) (okib.ca). The traditional language spoken by this community is the Nsyilxcən language, commonly referred to as the Okanagan language, categorized as critically endangered by UNESCO (Moseley, 2010). According to their estimates, the language is spoken by approximately 300 persons, with an average age of 59. The region has been a tourism destination for approximately one century, because of its dry, sunny climate in the summer. Since 2002, the band has owned and operated the Nk'Mip Tourism Resort and Winery, located on their reserve, which has led to economic development of the band, as well as the opportunity to showcase their heritage and history (okib.ca).

3.2 Cimbrian People in Giazza

The Cimbrian language is spoken by immigrants from southern Bavaria who moved to the mountains of northeastern Italy from the 11th to the 13th centuries (Bidese, 2020). The lords in the area called for the immigrants to chop the wood in the forests, in part because they were good carpenters (www.cimbri.it). The original area where people had Cimbrian ethnicity included the so-called *Tredici Comuni* (Thirteen Communities, northeast of Verona), *Sette Comuni* (Seven Communities, northwest of Vicenza), and the southeastern Trentino. Because the three areas were isolated from the city, they retained their traditional language, an Upper German variety derived from Southern Bavarian, called the Cimbrian language. Currently, this language is spoken only in three districts: Lusérn in Trentino, by elders in Giazza near Verona, and in Mezzaselva di Roana near Vicenza (Stringher, 2012). As a consequence, according to UNESCO's *Atlas of the World's Languages in Danger* (Moseley, 2010), the Cimbrian language is considered endangered, that is, children do not learn it as their mother tongue anymore, even though parents may speak it to them. Nonetheless, a clarification here is needed, as there is a significant difference among the three communities just mentioned. Nowadays, the community with the highest number of speakers is Lusérn, as, according to the latest census, in 2011, speakers numbered 238 out of 279 people living there, including children (Servizio Statistica della Provincia Autonoma di Trento, 2014). The situation is much more serious in Giazza, the area taken into consideration for the current study, where, in 2001, only 19 people (older than 65 years) out of a population of 131 could speak the Cimbrian language and another 24 (older than 45 years) could understand it. The young people could neither speak it nor understand it (Stringher, 2012).

During the 20th century, the fascist regime in Italy imposed the Italianization of the country so that the Cimbrian language was outlawed. Children were beaten on their hands (or worse) by teachers if they were caught speaking that Upper-German variety. After World War II, the economic miracle and the emancipation of the people living in Giazza resulted in a sudden endangerment of the language, because Cimbrian people were derided as being backward and ignorant. These are the two main reasons that have led to a rapid decline in use of the Cimbrian language in the past century (Stringher, 2012).

As already mentioned, the three linguistic enclaves are nestled at the foot of the Alps, making them a perfect destination for mountain lovers. For instance, the Lessinia area, the Alpine region near Verona, where Giazza is situated, is visited by hikers and people who practice mountain sports, as well as by winter lovers, even though in limited numbers (Ugolini & Costa, 2009). Giazza is the starting point for many hikes: from there hikers can reach not only the Foresta Demaniale di Giazza but also the Valle del Fraselle, as well as the more challenging Gruppo del Carega (2,259 m.a.s.l.) (www.giazza.it). However, because there are no hotels in the town (the only hotel closed some years ago) (www.giazza.it), visitors are still day-trippers; this could represent a problem for sustainability of the destination, as well as for promotion of the Cimbrian culture, that needs people to have the time to stop and visit the museum in town or take a cultural tour.

Even though it has been proved that the Cimbrian culture represents a potential for the development of a cultural product that could satisfy the needs of tourists (Ugolini & Costa, 2009), this little town still lacks a tourism offering specifically focused on it, even though visitors can find something related to the Cimbrian culture. The Curatorium Cimbricum Veronense, for instance, is a museum that offers an extensive explanation about the history of the area and the peculiarities of the Cimbrian culture. It even includes some in-depth explanations of the Cimbrian language. There are also events connected with the Cimbrian culture, such as the popular Festa del Fuoco (*Waur Ljetzan* in the traditional language, which can be translated into "fire feast"), held every year on the summer solstice. It is a pagan feast that celebrates the fire, water, and trees. Finally, in Bosco Chiesanuova, another Cimbrian municipality, not far from Giazza, the Film Festival della Lessinia is held at the end of the summer, which often presents films or other projects connected with the Cimbrian culture (www.giazza.it).

3.3 Methodology

This study was conducted on the basis of qualitative methodologies, that are extensively used in tourism research, especially when analyzing cultural and anthropological aspects of a population (Jennings, 2004; Richards & Munster, 2010). In particular, interviews have the main benefit of providing a holistic and detailed understanding of the phenomenon being studied (Jennings, 2004).

In total, 20 in-depth semi-structured interviews were conducted. Twelve persons were interviewed from the two local bands in British Columbia in the fall of 2017: both Indigenous and non-Indigenous people working at the Quaaout Lodge (n = 6) and Nk'Mip Resort (n = 6). In Giazza, eight persons, representing key players in the tourism industry and of Cimbrian-speaking heritage were interviewed in May 2018.

An important difference among the three cases is that in British Columbia, individuals representing and working for the two main Indian bands within a large tourism region were analyzed, while in Giazza the authors took the whole destination as an example. Giazza is a very small town, with a little more than 100 inhabitants (www.giazza.it), while, as regarding British Columbia, the focus was on two successful, Indigenous-owned and run companies in the Thompson-Okanagan tourism region, where community members were interviewed. The decision to include both Indigenous and non-Indigenous people in the sample followed the need to be as representative as possible of local residents working at the companies selected. In Giazza, interviewees included key players in the tourism field and in the documentation and study of the Cimbrian culture, together with restaurant owners.

The data collection was carried out by the first author. Key informant interviews were selected from the case studies for participation in the research. Interviews lasted up to an hour and were conducted in English in British Columbia and in Italian in Italy. Interviews were recorded with permission and transcribed for analysis using MaxQDA. Key emergent themes were identified and compared. Questions regarding language use in tourism and tourists' reactions were similar, although questions regarding language vitality and endangerment differed slightly between the cases in British Columbia and Italy because of marked differences in the history of the three case studies.

4. Findings

This section provides a summary of the general themes from transcription of the interviews. Themes discussed include how traditional language has influenced issues of discrimination and language preservation, tourism experiences, the sense of pride in and restoration of culture and community, and authenticity of the tourism product.

4.1 Discrimination and Language Preservation

As already mentioned, both Indigenous people in British Columbia and Cimbrian people in Giazza experienced such serious physical abuses and discrimination that they voluntarily stopped speaking their languages and passing them on to future generations. This was evident from interviews in the case studies.

Second, both interviewees in British Columbia and in Giazza very much agreed that language preservation is fundamental and urgent because they realize that their traditional languages are threatened. Most of them also associated the loss of their language with threatening their cultural heritage, since traditional languages are strongly tied to the traditional culture. They convey a particular meaning that could not be expressed otherwise. Respondents also commented that traditional languages stand for a distinctive cultural heritage, which represents an asset and a source of competitive advantage for the destination:

> I think we have something to offer that other wineries don't. When you come to visit our winery, it's not just another winery. . . . You're actually seeing some of our culture . . . we have a canoe with our sculptures, you get to see pictures and things like that. We integrate a bit of our culture into our winery.
>
> (Interviewee 3, winemaker at Nk'Mip Cellars)

> Especially on the reserves they want to hear the stories, they want to hear the words, they want to touch things that happened in the past to feel connected.
>
> (Interviewee 1, marketing manager, Desert Cultural Centre, Nk'Mip Resort)

"The words" are of particular importance in tourism experiences:

> It takes you back and ties to that time a little bit more than just saying 'it's a coyote, it's a deer. . . . It's not just a winter home, it's a *kekuli*. It's a *sәk'lép* [coyote].
>
> (Manager, Quaoout Lodge)

As illustrated in the previous quotations, traditional languages in British Columbia are used during the tours, especially to explain words tied to the culture and lifestyle of a place. The words for traditional dwellings, plants, or animals are often shared with tourists, and every staff member is also encouraged to use some simple words in the language (such as "hello", "thank you" and "goodbye"). However, in British Columbia, languages are also used to brand companies or products. The companies analyzed here adopted Indigenous names. *Quaaout* means "when the sun rays first touch the ground," and *Nk'Mip* means, "bottom land." Nk'Mip Cellars also uses Indigenous words related to the culture of the Okanagan people to brand their two premium wines. Tourists appreciated the use of traditional words and were happy to learn their pronunciation and meaning.

Cimbrian people also use their language in tourism, but in much more restricted domains. Interviewees recognize the value of using traditional words in tourist products. Visitors can, for instance, read bilingual signs in the town and in the path leading to the *carbonara*,[1] but it is rarely used in cultural tours, and sometimes for the branding of products.

4.2 Value of Language in Tourist Experiences

As we have already stated and proved in the course of this chapter, tourists are looking for unique experiences, and the distinctiveness of a culture thus represents an added value for the destination, which means it is easily marketable.

As interviewees point out, the Quaaout Lodge and the Nk'Mip Resort are not just "boxes, places to stay overnight" (manager, Quaaout Lodge) but are opportunities to live unforgettable and unique experiences, related to the place, its history and the history of its people. The Quaaout Lodge, for instance, was the promoter of a project connected with the construction of a canoe using traditional tools and accompanied by dances and religious practices. The ceremonies held during this project were attended by many visitors, and the videos uploaded on social media pages had thousands of views. This holds true for Giazza as well: the town and its surroundings are also not just any Alpine destination but can provide unique cultural experiences. Giazza has also become well known because of its rich history, not just because of its stunning landscape.

Because culture has a real differentiating value, people can actually make a living from their heritage. Findings from the case studies identified that working as a tour guide personally encouraged Indigenous people to go back and learn more about their culture and language. This happens not only on a personal level but also involves the whole company with activities such as encouraging the staff to attend language courses during paid staff time.

4.3 Sense of Pride in and Restoration of Culture and Community

The interest shown by tourists has led to a strengthened sense of pride among Indigenous peoples in their cultural heritage and a desire to do their best to preserve it, justified also by the economic benefits that are the result of culture and language preservation. The Indigenous people interviewed in British Columbia admitted that their jobs in the tourism industry actually were an incentive to learn more about their culture and languages, which eventually became part of a healing journey away from depression and drug and alcohol addiction.

Tourism also allowed the restoration of some ancient practices. At the Quaaout Lodge, for instance, a sweat lodge has been built and is regularly used, even though not all tourists can attend the highly spiritual purification ceremony. It is by invitation only. Other ceremonies are even more restricted, and only elders of the community can attend them.

In Giazza, the *Waur Ljetzan* traditional ceremony is becoming a popular event, with participants from outside the region. The tradition of coal production through the *carbonara*, typical of the Cimbrian people, has also been kept alive, in part thanks to the interests and appreciation shown by tourists:

> It's an added value. No one cooks with the coal he produces by himself. I'm proud of that.
>
> (Restaurant owner of the *osteria* in Giazza)

4.4 Authenticity of the Tourism Product

All of this, however, eventually leads to questions related to the authenticity of the culture displayed. In the beginning, there were many negative reactions to tourism, especially among Indigenous elders. They previously thought that tourism was a fake representation of the culture because it is sold to tourists. Indigenous people interviewed, to the contrary, believe sustainable tourism is part of their healing process, since, when the community is involved in the development, management, and promotion of tourist products, it also is able to pay attention to how the culture is displayed. They believe tourists nowadays are not interested in something fake. The problem is, however, that they often do not know what is authentic. For example, visitors in the Shuswap area expect to see totem poles, teepees, and all the headdresses and face paintings they saw in the movies. But this is not an authentic aspect of the Secwepemc heritage. Even in Giazza and its surroundings, where now the Cimbrian culture is getting "fashionable," the word *cimbro* in Italian is used to brand many food products. Whereas Cimbrian people have a long tradition of cheese production (now sold as *formaggio cimbro*), they did not traditionally produce beer or bake strudels or pizzas:

> Visitors can now find 'aperitivo cimbro, pizza cimbra' but these did not exist [formerly].
>
> (Museum director, Giazza)

On the other hand, tourism could be exactly the perfect platform to educate tourists about the differences among Indigenous people. And that is, again, one of the main reasons why Indigenous interviewees desired to learn about their past, their culture, and their language.

When specifically regarding the language, however, the issues of (in) authenticity take a slightly different path, maybe because learning a language requires real effort. Some people interviewed felt uncomfortable using words in the Indigenous or Cimbrian language, and some of them cannot even pronounce them. A restaurant owner in Giazza, even though he has Cimbrian origins and has lived in the town his whole life, said,

"I can´t promote the Cimbrian culture and language if I don´t speak it myself."

Non-Indigenous interviewees also had thoughts about it:

> All I have ever known is English. So . . . I can't say 'my grandmother used to speak this, my grandfather used to speak that and this is how important it is.
>
> (Manager, Quaoout Lodge)

Therefore, language is not used as a marketing gimmick or a folkloristic element but as a consequence of a personal pride enhanced by tourists' genuine interest.

Overall, some skeptical remarks were made as well. A critique coming from all three cases is the fact that for now language is relegated to museums or closed spaces, whereas it should be more within the landscape—for example, signs along the paths and hiking trails, as well as interpretation signs, should be bilingual, so that visitors immediately realize there is more than one language spoken there. Finally, some interviewees doubted that using a couple of words in connection with tourism could lead to proper language revitalization since language needs to be used daily for it to become more prominent.

What is certain for everyone interviewed, however, is that sustainable tourism helps bring awareness to language endangerment both inside and outside the community. Tourists in British Columbia leave with a better understanding of the differences between Indigenous people in North America and their numerous traditional languages, whereas in Giazza they get to know the Cimbrian culture and realize the language is still spoken nowadays. However, Indigenous interviewees argued that there is a need to educate people who live nearby, more than visitors. In Giazza, results show that a greater regional awareness and understanding of Cimbrian culture is important. People in the area need to realize the value of the Cimbrian culture, especially for the sustainability of the tourism industry in the region.

5. Discussion

Aligned with the literature, results showed that tourism is a vital tool for language preservation, especially if sustainably and responsibly managed. First, it plays an important role in developing an awareness of language endangerment, both inside and outside the community, which is seen as a step toward language preservation (Fishman, 1991). Second, it encourages people to stop being ashamed of their origins but rather to feel proud instead and learn more, so that the culture and the language do not fall into oblivion (Greathouse-Amador, 2005; Whitney-Squire, 2016). Language and cultural preservation are, therefore, not only the consequences of an enhanced pride and cultural renaissance but also of an economic benefit derived from it. The better preserved a culture is, the more distinctive it is and the more it represents an added value.

The results illustrate that there is a marked difference among the three cases analyzed, mainly because they have different histories and are now dealing with a different current situation. Indigenous people in British Columbia are aware that their culture represents a value and are proud to show it and to use and share traditional words in their unique languages with tourists (Whitney-Squire, 2016). However, people in Giazza are not at this point yet. They recognize that their culture is so distinct as to represent an asset, but, apart from some successful traditional feasts and practices, they still have not integrated their culture and language as extensively.

Both Indigenous people in British Columbia and Cimbrian people in Giazza experienced serious psychological and physical traumas because of their culture and language, even though to a different extent (Barten, 2015). This could have led to increased awareness and the rise of an Indigenous Renaissance. Moreover, Indigenous people of North America have long been portrayed in movies, and this has led to a rise of interests by non-Indigenous people and therefore eventually to the development of Indigenous tourism as a distinct field. There are now many tourists interested in learning more about Indigenous history, their heritage and their languages (Stirrup, 2013).

On the other hand, Cimbrian people still represent a minority of whom few people are aware. However, this does not mean tourists are not interested in their culture and language since results showed that Cimbrian culture is appreciated by tourists when authentically managed and displayed. In fact, Cimbrian people, even though they seem more reluctant, have an opportunity to develop tourism connected with the traditional aspects of their local culture, through food, traditional practices (e.g. the *carbonara*), and the Cimbrian language.

6. Conclusions

This chapter has shown that minority languages represent an asset for a tourism destination since they meet tourists' interest for unique and authentic cultural experiences. In the three case studies, languages were in fact not used as folkloristic elements to please tourists but as a consequence of personal pride enhanced by tourists' interest (Greathouse-Amador, 2005).

Tourism, if sustainably managed and promoted, has therefore positive impacts on the culture of a minority community (Soini & Birkeland, 2014; Kim et al., 2019). First, because tourists showed a real and authentic interest in these case study communities, the tourism field has the interest and funds to work for cultural preservation and achieve authentic culture revitalization (Whitney-Squire, 2016; Kelly-Holmes & Pietikäinen, 2014).

Even though there are some doubts about a proper revitalization achieved through the use of just a couple of words in traditional languages in connection with tourism, the industry definitely helps bring awareness both inside and outside the community. On one hand, tourists leave with a better understanding of traditional languages and their degree of endangerment. On the other hand, these same members of the community often feel motivated to learn more after they realize how much their culture and language are valued, which will, in turn, reinforce the authenticity of the experience.

It is hoped this chapter is a stepping-stone in the analysis of language preservation through tourism, especially among minority groups in British Columbia and Italy. Further research is therefore needed, not only regarding in-depth analysis of other communities in British Columbia and Italy but also to conduct a broader quantitative analysis of tourists' perception of and interest in minority language use in tourism experiences.

Note

1. Charcoal pile, a traditional way to produce coal. The tradition has been revitalized by the restaurant owner of the *osteria* in the main town square.

References

The Aboriginal Tourism Association of British Columbia (2015). *The next phase: 2012–2017 — A five-year strategy for aboriginal cultural tourism in British Columbia*. Vancouver: The Aboriginal Tourism Association of British Columbia.

Aboriginal Tourism Association of Canada. (2016). *National guidelines: Aboriginal cultural experiences*. Vancouver, BC: Aboriginal Tourism Association of Canada.

Archer, B., Cooper, C., & Ruhanen, L. (2005). The positive and negative impacts of tourism. In W. F. Theobald (Ed.), *Global tourism* (3rd ed., pp. 79–102). Burlington, MA: Elsevier.

Barten, U. (2015). What's in a name? Peoples, minorities, indigenous peoples, tribal groups and nations. *Journal on Ethnopolitics and Minority Issues in Europe*, *14*(1), 1–25.

BBC news. (2010, January 1). *Aboriginal Canadians divided over Vancouver Olympics*. Retrieved October 24, 2017, from http://news.bbc.co.uk/2/hi/8426055.stm

Bidese, E. (2020). Italy I: Cimbrian. In H. Boas, A. Deumert, L. M. Louden, & P. Maitz (Eds.), *Varieties of German worldwide*. Oxford: Oxford University Press.

Burden, M. (2007). Museums and the intangible heritage: The case study of the Afrikaans language museum. *International Journal of Intangible Heritage*, 2, 81–91.

Butler, R., & Hinch, T. (Eds.). (2007). *Tourism and indigenous peoples: Issues and implications*. Oxford, UK: Routledge.

Capotorti, F. (1977). *Study on the rights of persons belonging to ethnic, religious and linguistic minorities*. New York: United Nations.

Cobo, J. R. M. (1986). *Study of the problem of discrimination against indigenous populations*. Geneva: United Nations.

Cohen, E. (1988). Authenticity and commoditization in tourism. *Annals of Tourism Research*, *15*(3), 371–386.

Crawford, J. (1995). Endangered native American languages: What is to be done, and why? *The Bilingual Research Journal*, *19*(1), 17–38.

Duchêne, A., & Heller, M. (Eds.). (2012). *Language in late capitalism: Pride and profit*. London: Routledge.

Fishman, J. A. (1991). *Reversing language shift: Theoretical and empirical foundations of assistance to threatened languages*. Clevedon and Philadelphia: Multilingual Matters.

Fleming, W. C. (2006). Myths and stereotypes about native Americans. *Phi Delta Kappan Magazine, 88*(3), 213–217. doi:10.1177/003172170608800319

Greathouse-Amador, L. M. (2005). Tourism and policy in preserving minority languages and culture: The Cuetzalan experience. *Review of Policy Research*, *22*(1), 49–58.

Heinrich, P. (2010). Language choices at Naha Airport. *Japanese Studies*, *30*(3), 343–358. doi:10.1080/10371397.2010.518599

Heller, M. (2003). Globalization, the new economy, and the commodification of language and identity. *Journal of Sociolinguistics*, *7*(4), 473–492. Retrieved July 3, 2019, from https://okib.ca/

Indigenous Tourism BC. (2018). *ITBC "the next phase" tourism performance audit report 2012–2017*. Vancouver: Indigenous Tourism BC. Retrieved February 3, 2020, from www.indigenousbc. com/drive/uploads/2018/10/REPORT-ITBC-Audit-2012–2017_FINAL.pdf

Jennings, G. R. (2004). Interviewing: A focus on qualitative techniques. In B. Ritchie, P. Burns, & C. Palmer (Eds.), *Tourism research methods: Integrating theory with practice* (pp. 99–117). Wallingford: CAB Publishing.

Kelly-Holmes, H., & Pietikäinen, S. (2014). Commodifying Sámi culture in an indigenous tourism site. *Journal of Sociolinguistics*, *18*(4), 518–538. doi:10.1111/josl.12092

Kelly-Holmes, H., & Pietikäinen, S. (2016). Language: A challenging resource in a museum of Sámi culture. *Scandinavian Journal of Hospitality and Tourism, 16*(1), 24–41. doi:10.1080/15022250.2015.1058186

Kim, J.-H., Ritchie, J. R., & McCormick, B. (2012). Development of a scale to measure memorable tourism experiences. *Journal of Travel Research*, *51*(1), 12–25.

Kim, S., Whitford, M., & Arcodia, C. (2019). Development of intangible cultural heritage as a sustainable tourism resource: The intangible cultural heritage practitioners' perspectives. *Journal of Heritage Tourism*, *153*(3), 1–14. https://doi.org/10.1080/1743873X.2018.1561703.

Knudsen, D. C., Rickly, J. M., & Vidon, E. S. (2016). The fantasy of authenticity: Touring with Lacan. *Annals of Tourism Research*, *58*, 33–45.

Kolar, T., & Zabkar, V. (2010). A consumer-based model of authenticity: An oxymoron or the foundation of cultural heritage marketing? *Tourism Management*, *31*(2010), 652–664.

Krauss, M. (1992). The world's languages in crisis. *Language*, *68*(1), 4–10.

MacCannell, D. (1973). Staged authenticity: Arrangements of social space in tourist settings. *American Journal of Sociology*, *79*(3), 589–603.

Maffi, L. (2007). Biocultural diversity and sustainability. In J. Pretty, A. Ball, T. Benton, J. Guivant, D. R. Lee, D. Orr, . . . Ward, H. (Eds.), *The SAGE handbook of environment and society* (pp. 267–277). London: SAGE.

Mason, C. W. (2014). *Spirits of the Rockies: Reasserting an indigenous presence in Banff National Park*. Toronto: University of Toronto.

Moriarty, M. (2014). Contesting language ideologies in the linguistic landscape of an Irish tourist town. *International Journal of Bilingualism*, *18*(5), 464–477.

Moseley, C. (2010). *Atlas of the world's languages in danger*. Paris: UNESCO Publishing.

Mufwene, S. S. (2002). Colonisation, globalisation, and the future of languages in the twenty-first century. *MOST Journal on Multicultural Societies*, *4*(2), 162–193.

Partridge, C. (2011). Residential schools: The intergenerational impacts on Aboriginal peoples. *Native Social Work Journal*, *7*, 33–62.

Pietikäinen, S., Kelly-Holmes, H., & Peer, H. (2011). The local political economy of languages in a Sámi tourism destination authenticity and mobility in the labelling of souvenirs. *Journal of Sociolinguistics*, *15*(3), 323–346. doi:10.1111/j.1467-9841.2011.00489.x

Province of British Columbia. (2011). *Super natural British Columbia: An introduction to our brand*. Vancouver: Province of British Columbia.

Qilou, Z., Zhang, J., Zhang, H., & Ma, J. (2015). A structural model of host authenticity. *Annals of Tourism Research*, *55*(2015), 28–45.

Richards, G., & Munsters, W. (2010). *Cultural tourism research methods*. Wallingford: CABI.

Ritchie, B. W., Cooper, C. P., & Carr, N. (2003). *Managing educational tourism*. Clevedon: Channel View Publications.

Servizio Statistica della Provincia Autonoma di Trento. (2014). *Rilevazione sulla consistenza e la dislocazione territoriale degli appartenenti alle popolazioni di lingua ladina, mòchena e cimbra*. 15° Censimento generale della popolazione e delle abitazioni—dati definitivi, Trento.

Soini, K., & Birkeland, I. (2014). Exploring the scientific discourse on cultural sustainability. *Geoforum*, *51*, 213–223.

Statistics Canada. (2016). *Focus on geography, 2016 census British Columbia*. Ottawa: Statistics Canada.

Stirrup, D. (2013). Introduction. In J. Mackay & D. Stirrup (Eds.), *Tribal fantasies: Native Americans in the European imaginary, 1900–2010* (pp. 1–19). New York: Palgrave Macmillan.

Stringher, A. (2012). *Censimento dei parlanti Cimbro nell'isola linguistica di Giazza: Consistenza della parlata tedesca dei tredici comuni veronesi dal XVII al XX secolo*. Selva di Progno, VR: Comune di Selva di Progno.

Ugolini, M., & Costa, K. (2009). I cimbri come fattore di attrattiva turistica? *Quaderni di lingue e letterature*, *34*(Supp. l), 113–133.

UNESCO Ad Hoc Expert Group on Endangered Languages. (2003). *Language vitality and endangerment*. Paris: UNESCO.

Whitney-Squire, K. (2016). Sustaining local language relationships through Indigenous community-based tourism initiatives. *Journal of Sustainable Tourism*, *24*(8/9), 1156–1176.

www.cimbri.it. Retrieved May 3, 2018.
www.giazza.it/index.htm. Retrieved May 2, 2018.
www.lslib.com/. Retrieved July 3, 2019.
Young, B. (2015). "Killing the Indian in the child": Death, cruelty and subject-formation in the Canadian Indian residential school system. *A Journal for the Interdisciplinary Study of Literature*, *48*(4), 63–76.
Zhu, Y. (2012). Performing heritage: Rethinking authenticity in tourism. *Annals of Tourism Research*, *39*(3), 1495–1513.

11 Spontaneous Tourism and Sustainable Development

The Evolution of the City of Naples

Fabiana Sciarelli, Valentina Della Corte, and Giovanna Del Gaudio

1. Introduction

The aim of this chapter is to develop an understanding of sustainable destination development with respect to metropolitan cities that in recent years have shown spontaneous tourist growth.

The managerial approach to sustainable tourism development implies an overlapping logic between the demand and the supply perspective and the identification of a unit of analysis that allows establishment of a series of links between the two dimensions, that is, destination. A destination, in fact, can be defined as an integrated system of attracting factors (natural, cultural, enogastronomic [food and wine sector], etc.) and structures for the promotion and management of hospitality (hospitality, institutions, transport, catering, etc.) Destination can be analyzed according to the 6As model (Access, Accommodations, Attractions, Amenities, Ancillary services, Assemblage) (Della Corte, 2013).

This chapter draws upon a literature review, which sheds light on the overlapping perspectives on the topic and explores what factors enhance competitiveness within the urban context according to a sustainable tourism development's perspective as well as the existing relationships among actors of the tourism industry, local citizens, tourists, and other stakeholders of the ecosystem (the whole of actors involved in the destination).

From a methodological point of view, this chapter addresses the issue of sustainable destination development with respect to metropolitan cities, particularly by examining the case of the city of Naples in Italy.

The strength of the investigation lies in the analysis of a city that has witnessed spontaneous tourist growth and therefore offers a different perspective from previous contributions on the topic of sustainable destination development.

2. Material and Methods

To discuss our research question, the study is organized as follows: we first provide the essentials of the conceptual and theoretical background of our proposal by illustrating the concepts of sustainable tourism and spontaneous tourism. Subsequently, we present Naples' case study, highlighting elements useful to the subsequent discussion where we propose our view of spontaneous and sustainable tourism development. Finally, we highlight the main managerial and research implications.

To illustrate our theoretical background, we first briefly introduce the concepts of sustainable tourism and spontaneous tourism as well as their interconnections. The research methodology adopts a case study analysis and analyzes the relative qualitative results. Furthermore, the case study methodology is in line with the exploratory nature of this chapter since it is able to capture contextual richness and complexity of research issues (Yin, 2003), as well as understanding the social structures (Riege, 2003).

To capture these accurate reflections on sustainable destination development with respect to the city of Naples, in-depth interviews with pivotal actors (presidents of the main tourism associations, top managers of the Naples Convention Bureau) and other members were conducted following a predesigned protocol (Yin, 2003). Indeed, the interviews used a snowball method that allows information to be caught from both central actors and peripheral ones in order to obtain a more holistic view of who manages and how he manages the tourism sector according to a sustainable perspective.

3. Theoretical Background

3.1 Meaning of Sustainable Tourism

Since 1987, the year in which the Brundtland Report (WCED, 1987) was published, sustainable development has become a fuzzy concept (Boluk, Cavaliere & Higgins-Desbiolles, 2017), dealing with different research topics within the tourism industry (i.e. governance of sustainable destinations, different pillars of sustainable tourism, models for sustainable tourism development, etc.).

According to the Brundtland Report:

> Sustainable development is not a fixed state of harmony, but rather a process of change in which the exploitation of resources, the direction of the investments, the orientation of technological development, and institutional change are made consistent with future as well as present needs.
>
> (WCED, 1987 p. 9)

Despite the numerous definitions (Ritchie & Crouch, 2003; Hunter, 1997) of sustainable tourism, the one provided in the Brundtland Report remains one of the most current, given also the fact that subsequent studies have then recalled the main elements enclosed in it.

As a matter of fact, this is a recurring term always applicable and valid with reference to two relevant issues:

1) "temporality", namely, there should be a continuous process that integrates both quantitative and qualitative "conservation" of existing resources ("unlimited time" for the present and the future);
2) "overlapping logic of more dimensions", in that it overcomes the sectorial sustainability linked exclusively to the environmental ecosystem, while invoking a balance between other dimensions, such as the economic and the socio-cultural one.

However, subsequent studies have also incorporated other elements in their analyses/analytical frameworks, such as the integration and co-planning logics of the tourism offer, in harmony with processes of protection and enhancement within territorial resources (WTO, 2002; Romei, 2009); the tourist-resident relationship; and the importance of preserving and respecting the latter, that is, the local community from the impacts of tourism processes (Swarbrooke, 1999; Lim & Cooper, 2009).

In light of these reflections, which are connected with the role of local community, this chapter explores the role of local community and other involved stakeholders with regard to spontaneous tourism. In the expression "spontaneous tourism", "spontaneous" relates to an overlapping perspective between both the supply and the demand side.

As a matter of fact, from the supply side, destinations that have experienced tourist growth through spontaneous development (Pechlaner, Raich, Beritelli, d'Angella, De Carlo & Sainaghi, 2010) have not been established on the basis of the destination management organization (DMO). This means that there are no strategic programs and guidelines followed by tourist firms, local communities, and other stakeholders in the fields of either tourism growth or sustainable tourism development. Hence, tourists are not driven by the traditional pull and push factors (Martini, 2005). These latter are actions planned by the destination regions that motivate the tourist's choice of the destination or one destination over another. The term "spontaneous" means causal since it is determined by tourist choices apart from the level of organization at a destination level and the lack of planned strategic and marketing plans. Thus, tourists spontaneously choose where to travel.

Consequently, tourism development is spontaneous and, within this framework, tourist firms, public bodies, and other private companies have organized their own policies without having a clear vision of the strategic tourism development of the destination (Saraniemi & Kylänen, 2011). In most cases, local organizations adopt strategies as a response to local environmental changes and needs.

From the demand side, the phenomenon of spontaneous tourism derives from the lack of strategic and marketing policies within the network, which can better attract tourists according to their vocation of destination (De Rosa & Salvati, 2016; Perales, 2002).

In this context, actions concerning sustainable tourism development are the result of single actors (i.e. local communities, tourism firms, transportation firms, etc.) that plan in an autonomous way the best practices for social, environmental, and economic sustainable growth (Briassoulis, 2002; Tremblay, 2000).

Notwithstanding the numerous efforts made by different actors, at a certain point, the phenomenon of spontaneous tourism has to deal with either top-down or bottom-up logics (Rodriguez, Williams & Hall, 2014). In point of fact, the politics of spontaneous tourism and, consequently, the formulation of single development practices may also lead to asymmetric information, massive fragmentation, and effort duplication. Furthermore, individual strategies are not always sustainable, since the single actors cannot own the specific and holistic competences to plan appropriate/adequate actions.

Therefore, despite the individual capabilities that local organizations and entities may have or control, the lack of a specific territorial strategic orientation can prevent grasping all the possible opportunities offered by the market.

3.2 Planning Sustainable Tourism Development

A possible remark may be that, in view of spontaneous processes, a systemic approach should develop anyway. In such a case, it usually happens that a group of firms that share a common strategic vision start cooperating and then aggregating other actors. This is the most successful path of spontaneous tourism, even if it is important then to understand how representative the aggregation is of the whole territory.

Such considerations have led many scholars (Garcia, Luengo, Sáez, Lopez & Herrera, 2012; Della Corte, Sciarelli & Del Gaudio, 2018; Heslinga, Groote & Vanclay, 2019) to state that a double approach is often necessary, and it should combine both bottom-up and top-down processes: the former expresses entrepreneurial initiative, and the latter provides a wide, open perspective in territorial planning and investing. This is relevant for multiple reasons—namely, the involvement of other institutional partners whose role is pivotal (airports, ports, and train station companies), especially with regard to planned and sustainable development, in terms of infrastructures, destination image, and overall destination marketing.

From this point of view, strategic planning is crucial for sustainable tourism development happening through spontaneous processes (Liburd & Edwards, 2010; Della Corte & Sciarelli, 2013) and requires the active role of a governance actor in the immediate next stage of unplanned tourism.

The planning phases refer to the formulation of activities that have to be (Liburd & Edwards, 2010; Della Corte & Sciarelli, 2013):

- Goal-oriented: in this case, the role that tourism covers within the local community should be clear to the governance bodies, in order to achieve objectives that can satisfy the interests of different stakeholder systems;
- Integrative: correct strategic programming, according to a sustainable perspective, should take into account all the elements that contribute to form the destination (6As model);
- Market-driven: attractiveness is an essential feature when defining a destination. Nowadays, destinations compete on a global scale, and it is important that they are attractive to targeted tourists;
- Resource-driven: the activities of the destination's pivot organs must be oriented toward balancing the strengthening of destination strategic assets and the protection of resources in the fruition phase of the territory. The resources are, in fact, strategic factors in the process of sustainable tourism development, although they may generate conflicts between the pillars of sustainability (i.e. between the economic dimension and the social and ecological dimensions);
- Of consultative nature: the planning of sustainable activities should be the result of a co-creation process activated by the governance bodies for the formulation of value offers. In this process, tourists, local communities, and other players represent fundamental inputs, within the supply chain, for understanding of the different interests underlying the construction of the sustainability concept;
- Systemic: the DMO has to formulate best practices shareable with the governance bodies of other destinations. Therefore, a systemic perspective that goes beyond the territorial boundaries is needed to build and strengthen the contents of sustainable tourism development.

3.3 Theoretical Framework

Our theoretical framework emerges from these reflections. Figure 11.1 shows how spontaneous tourism can lead to planned and systemic strategies also in the sustainability field. In fact, in the first stage, where tourism flows are spontaneous, development strategies would mirror the single voice of tourism firms. In the following phase, networking actions would subsequently be necessary shaping a humanistic perspective for the tourism industry to value previous efforts made by the single actors.

Therefore, the suggested step from "spontaneous" to "planned" tourism (Park, Rene, Choi & Chiu, 2008) requires the capability of the governance actors to employ all the positive and identity-based sustainable activities enacted by the different actors of the ecosystem.

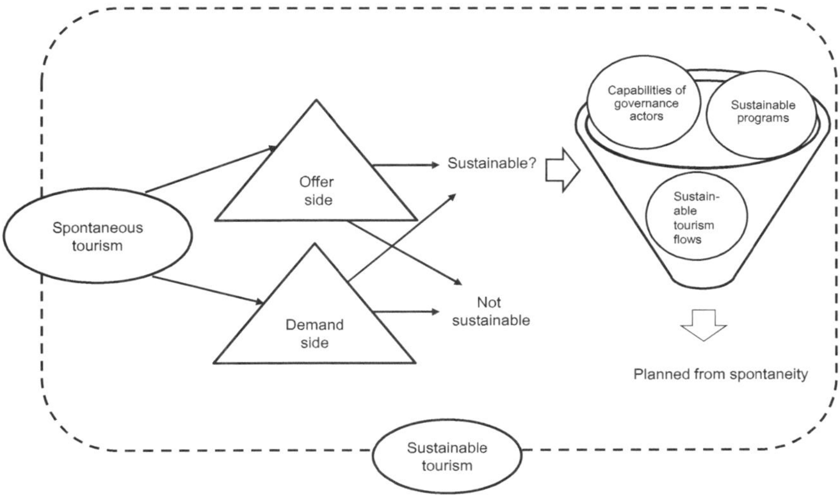

Figure 11.1 Spontaneity and sustainability of tourism

As can be seen from the model in Figure 11.1, spontaneity on both the demand and offer sides may generate positive effects on sustainable tourism practices. On the offer side, tourism firms and the other actors involved in the tourism industry should strive to (Lehtonen, 2004; Dwyer, 2005; Paniccia & Leoni, 2019):

- Adopt practices that avoid and reduce the exploitation of environmental resources as well as employ resources differently, thus having a minor impact on the environment;
- Create positive social impact, for example, by hiring local people, fostering local resources and donating money to local volunteer associations;
- Create added economic value and enhance revenues connected with foreign markets so as to distribute value to social stakeholders and internal human resources throughout the territory.

On the demand side, tourists can engage in sustainable practices in different ways, by:

- Favoring natural destinations and attractions in accordance with their interest and preserving these resources in a proactive way (Dolnicar, 2004; Zhang, Wang, Sun, J. & Yu, 2019);
- Orienting their decisional behavior toward sustainable choices and practices (i.e. when choosing modes of transport, accommodation, etc. Hanna & Adams, 2019);

- Co-creating sustainable experiences, suggesting ideas for greener activities, etc. (Jovicic, 2019).

However, spontaneous tourism may also generate practices and routines that are far from being sustainable, as the arrow on demand side in Figure 11.1 shows. Hence, leading actors in the governance of the territory, also through a planning process, favor the translations of the capabilities and best practices that have been developed in the spontaneous process into a more aware explicated and guided vision of a concrete process of sustainable tourism. This approach can take to a positive flow of actions that can account for the spontaneous representation of the local identity, thus naturally expressed by the actors involved in the process.

4. Case Study: Tourism in Naples Between Spontaneity and Sustainability

4.1 Context: Italy and Campania

Over the past 20 years, tourism has experienced an extraordinary worldwide expansion, supported by lower transport costs and rising income levels. This has also affected emerging economies, resulting in a pool of potential travelers which has enormously widened. This increase in demand has been accompanied by the affirmation of new destinations, which have attracted a growing number of tourists.

Italy is among the countries with the oldest vocation for tourism. At the beginning of the 1980s, when tourism was still limited to a few international destinations, it was second only to the United States in terms of impact on global tourist spending. Italy also boasts an artistic heritage and natural resources with few equals in the world: with 54 of the 1,092 UNESCO sites, Italy is the lead country for having places recognized as world heritage sites.

In Italy, all tourist activities account for more than 5 percent of GDP and more than 6 percent of the country's employees (Banca d'Italia, 2018), an economic weight comparable with that of Spain and higher than that of France and Germany.

In the face of these global trends, Italy's market share—like that of other mature tourist destinations—has inevitably contracted: from 7 percent of the world tourist spending in the first half of the 1990s to 3.4 percent in 2017 (Banca d'Italia, 2018). The decline, although in part physiological, was more intense for our country than for the main European competitors.

Between the late 1990s and the beginning of this decade, expenditures by foreign tourists in Italy grew substantially, no less than the overall spending of international tourists and the potential demand expressed by the countries of origin of traditional tourism for the Italian country.

In the same period, the balance of payments on travel—while remaining the only item historically active—has reduced by more than half a point of GDP.

Since 2010 alone, there have been some signs of recovery, partly favored by an improvement in price competitiveness (Banca d'Italia, 2018) and partly because of the rise of geopolitical tensions, which has discouraged travels to several competing countries that were at greater risk of terrorist attacks. Foreign spending in Italy has started to increase again at a sustained pace (4.3 percent per year on average, compared with 0.8 percent in the previous decade), significantly reducing the growth gap related to the potential demand for tourist services, which remained however negative. At the same time, the balance of payments related to travel balance returned to growth, amounting to 0.9 percent of GDP in 2017.

Among the distinctive features of this recovery is certainly a renewed interest of foreign tourists for vacation, and particularly for cultural holidays in Italy. In the period 2010–2017, international traveler spending on cultural holidays grew by almost 9 percent a year (Banca d'Italia, 2018). As for the provinces preferred by tourists, the Italian province with the largest revenue inflow in 2017 was Rome (6,743 mln), which recorded an increase of 20.3 percent compared with the previous year. Venice and Naples also showed a significant increase (+19.4% and +17.8%, respectively), while for Milan and Florence revenues were in decline (–2.4 and –6.3%, respectively) (ROMA, 2018).

Campania is the first region in Southern Italy for tourist flows: in the three-year period 2016–2018, arrivals grew at an average annual rate of 5.17 percent, significantly higher than the same value for the entire country, which stood at 3.4 percent.

In 2018, in particular—a year for which data are still provisional—arrivals and presences increased by 7.7 percent and 3.3 percent, respectively, which in absolute value correspond approximately to 6.075 million and 21.132 million (Becheri, Micera & Morbillo, 2018). In addition, in 2017, tourist spending in Campania amounted to 6.041 million euros, ranking sixth after Lombardy (13.6%), Lazio (11.5%), Tuscany (11.5%), Veneto (9.8%), and Emilia Romagna (8%).

In 2017, Campania recorded an annual increase in beds of 16 percent, reaching a total of 232,592, equally distributed between hotel and extra-hotel facilities, with a weight in relation to the national total that rose from 4 percent to about 5 percent.

Moreover, from 2015 to 2018, the local units of enterprises in the entire tourist system in Campania increased overall by approximately 10.6 percent, reaching 65,215 units. This, consequently, generated a growth of 23.7 percent in the number of employees, from 140,235 to 173,456.

A more detailed analysis shows how—despite the availability of six UNESCO sites out of a total of 54, 18 Blue Flag beaches out of 175

(10.29%), and 17 municipalities with at least one thermal establishment out of a total of 190 (8.95%)—in 2017, the region attracted tourist flows whose incidence on the national total was lower than that of the relative structures.

In detail, the following numbers were recorded in the region: of the tourist national movement, about 3.7 percent was directed to cities of historical and artistic interest, 6.31 percent to seaside resorts and 6.6 percent to spa resorts.

On the other hand, the positive performances of recent years have only slightly reversed the gradual contraction of the region's market share compared with the national total. The contraction started in 2001, the year in which the Campania tourism movement had an incidence of 6.1 percent, which decreased to 4.8 percent in 2015, to then went up again in 2017 to 4.9 percent. Added to this is the high seasonality, which in 2017 reached the lowest point in February (442,614 presences), the peak point in August (3,792,414 presences), to then fall again in November (689,370 presences).

Furthermore, the major concentration of tourist flows and beds in the coastal areas in the provinces of Naples (64.4%) and Salerno (29.5%)—a phenomenon common to all countries—assumes particularly anomalous connotations in Campania, generating on the one hand a risk of massification in these territories, and on the other excessive marginalization of internal areas. These critical issues could also be attributed to a regional governance that is struggling to overcome them and that today presents numerous elements of complexity and a risk of overlapping competences.

The data reported so far come officially from ISTAT, but there is at least an equally significant part of the phenomenon that is not detected, both because there is no obligation in this sense, and because it is largely not declared, giving rise to the phenomenon of the submerged data. (ISTAT data, 2017). Particularly, this refers to qualitative data such as the planned actions for sustainable development, whether and to what extent tourist firms collaborate for destination marketing and management and whether and to what extent tourists perceive the tourism offer as sustainable.

4.2 Naples: Data, Trend, Result

The growth of cultural tourism in Italy is already a consolidated trend. This has driven another sector to become fundamental in the cities of art; namely, the extra hotel circuit, B&B, and holiday homes, which today represent more than half of the available beds, increased overall by a total of 196,000 units. This phenomenon has left ample room for growth to new small-tourism entrepreneurs, who have considerably increased tourist services to medium-level spending by offering products with a good price-to-quality ratio. However, the classical receptive activities have also enjoyed this growth, and, as a matter of fact, from 2010 to

2018 they have gone up by 32,000 units, a percentage growth of 126 points (Mazzone, 2019).

Naples is among the top five cultural tourism locations that have recorded the best performances. Between 2010 and 2018, it was the second-highest Italian city in cultural tourism increase—a trend in line with the national one. The record belongs to Matera (European Capital of Culture in 2019), with a boom of 176% in tourism. Naples occupies second place in the ranking of cities having the largest tourist growth, with an increase of 108.7% compared with 2010 (Assoturismo-Cst data for Confesercent, 2019). The city today enjoys a revival of its image, with its many cultural attractions that guarantee the satisfaction of a diversified and global demand.

To confirm all of this, the data provided by the municipality of Naples (2012–2017) show an average annual growth in arrivals of 9% and presences of 8%, and record an average number of days per tourist of two in 2017.

In Naples, foreign tourists spend, on average, 124.9 euros per capita a day; that is below the Italian average of 129 euros and is distinctly less than in other major Italian cities of art (Milan, 155.1,; Florence, 153.7; Rome, 142.7; Venice, 138.9; Turin, 126.4).

Furthermore, with reference to the 2017–2018 season, Naples had the highest increase in tourists, equal to 13.3% (+11.2% Italians, +15% foreigners), with Palermo (with +11.9% for Italians) and Perugia (+15.7% of foreigners) just above (Stylo, 2019).

The Lonely Planet website indicates that Naples is the coolest city in Italy for four main reasons: 1) nightlife; 2) street art, highlighting the murals that started to dress public housing; 3) the high number of vintage shops; and 4) domestic receptivity that hints at the typical hospitality of the Neapolitan people.

Further, this designation improves the international image of Naples. However, being cool is typically seen as a short- or medium-term phenomenon, especially if not managed and made sustainable in the long run. This could yet be achieved through specific actions that transform fashion into a stable value.

These specific actions, having to do largely with elements that are not measured but that involve the human sphere, the environmental beauty, and the consequent quality of Neapolitan life, are unquestionably linked to humanistic management. This is the only management style capable of governing a complex tourist experience in which the local population becomes a central player in the local tourism process.

4.3 Spontaneous Tourism in Naples

According to Gennaro Biondi, 2016, this remarkable increase finds its *raison d'être* in at least three exogenous motivations. First, the tendential growth of international tourism now shows an annual growth of more than 5%. Of this greater demand, Naples (like all of South Italy) intercepts a much smaller percentage than the major European tourist cities (as well as Campania, e.g. with respect to Catalonia). Second is the destination changes "forced" by the new geo-politics, which are being defined in the Mediterranean Basin. In fact, the southern front from Maghreb to Machrek was destabilized first by the "Arab Spring", then by regional conflicts, and ultimately by terrorism, to the point that in some countries (e.g. Egypt and Tunisia) tourism has more than halved. Third, there is an increasing propensity for holiday-makers to benefit from tour operators interested in economic affordability and traveler safety. This has coincided in Naples with the policy of many hoteliers, who have sought competitiveness through price control rather than on improvement and expansion of services.

These can be seen as cyclical phenomena that, as the history of the sector shows, may have a short life and therefore lead to tourism spontaneity that ultimately is not sustainable.

To these phenomena may certainly be added the spontaneous action of the Neapolitan population that has been creating a fabric of tourist micro-entrepreneurship, which brings immediate advantages, chiefly, safeguarding of the territory. This is the case in Quartieri Spagnoli, where the alleys have always been considered to be places of crime and therefore inaccessible to people. Today, thanks to the numerous houses for rent, B&Bs, small restaurants, and visits offered by youth cooperatives, these alleyways have become absolutely accessible, livable, and safe, providing advantages for residents also.

Spontaneous micro-entrepreneurship also leads to a greater need for environmental and urban care. Because tourism is ascertained as the industrial engine for the Neapolitan territory, it is the entrepreneurs who should demand and handle the cleaning and maintenance of the places.

In addition, the joint interest of local residents and entrepreneurs brings a level of affection to the territory that is not yet fully discernible.

A consistent demand, created by the positive conjunctures and the rising supply, can therefore be singled out. Furthermore, a positive contribution to the image of the city was certainly given by the Film Commission Campania, which fostered the overcoming of a period in which Naples was solely linked to crime and waste problems. The adamant work of its director managed to bring home some interesting projects such as the fictions "Sirene", "I Bastardi di Pizzofalcone", and also "Gomorra", which have not only improved the image of the city but have certainly amplified its charm.

In sum, the problem today lies in how to render the economic growth structural, while identifying forms of intervention on a local and metropolitan scale that give answers to the new features characterizing the tourist demand. Particular attention should be paid to that segment that includes the "do-it-yourself traveller", that is, the tourist who seeks "an experience to live" and not just a place to visit.

All this, consequently, imposes a substantial change of perspective, starting from the hypothesis that competitiveness in the sector is no longer between single localities with their material heritage but rather between economic-territorial systems, whose organization should include adequate material and immaterial infrastructural policies as well as complementary services that create value (Biondi, 2016).

Faced with these processes, companies in the tourism sector are showing growing resourcefulness, a greater systemic vision and a firm will to favor the takeoff of the area. The constitution of the convention bureau (CB) is an expression of this process in the meetings, events and congresses sector.

The latter consists of a network of private operators that represent the top players in the Neapolitan meeting industry. The mission of the CB is to enhance the exclusivity of Naples, promoting the city for the organization of events, congresses, incentive trips, and team building. The Convention Bureau Napoli attracts conference organizers from the national and international scene, who are interested in organizing events in Naples. The task of the CB is to facilitate and coordinate the relationship with the institutions and the operators within the sector.

This complex, spontaneous process, regulated by a state of freedom, in line with humanistic management, shows how social cooperation in territories such as the Neapolitan one rewards more than individual competition.

What remains to be understood is the link between spontaneous development processes, in a systemic and humanistic perspective, and the assurance of a sustainable approach in terms of destination management.

4.4 Sustainable Tourism in Naples

Italy historically presents a manifest deficit in the capacity of programming and coordination among the different levels of government and between the latter and stakeholders in the sector, with negative effects on the country's ability to fully grasp the potential of the sector (Sciarelli, 2007; Della Corte, 2013).

This weakness reflects both the complexity of articulation of institutional competences, partly attributable to the progressive weakening of the central government's role, and a wavering trend of the political choices in the past quarter of a century. The residual role of coordination entrusted to the central government, for example, has not always received the attention that a sector of such importance would have deserved, nor has it had a stable institutional position with respect to the same subject, since the specific ministry of tourism was abolished in the 1990s.

The Strategic Tourism Development Plan (PST) 2017–22 was approved with Government Act n. 372 of 17/2/2017, at the end of a journey started two years earlier with the "General States of Tourism" of Pietrarsa. The methodology introduced is innovative with respect to the past, being inspired by an open and participatory method, which provides for the systematic comparison among all the institutions involved and between the latter and the sector operators, in line with OECD (Organisation for Economic Co-operation and Development) recommendations.

The main purpose of the plan is to relaunch the tourist attractiveness of Italy, drawing inspiration from three transversal principles: sustainability (environmental, mobility, full use of heritage, enhancement of identities), innovation (of the organizational process and the product, with particular emphasis on digitalization), and accessibility (access modality to places and possibilities for tourist use). The document is developed around four fundamental objectives: 1) the diversification of the tourist offer, 2) the increase in competitiveness of the system, 3) the development of effective and innovative marketing, and 4) the implementation of efficient and participatory governance. Each principle is divided into specific objectives and lines of intervention.

In January 2018, the first PST Implementation Program for the 2017–18 biennium was approved. Fifty initiatives connected to strategies and objectives of the PST were identified, and approximately 600 million euros were allocated to them. At the same time, a surveillance and evaluation system was carried out to measure the effectiveness and efficiency of the tourism development strategies and, in particular, the actions implemented in the program.

In any case, this is a first step along a path that requires consistency and determination in the upcoming years. This should be done also in conjunction with the new asset of ministerial attributions, through the implementation of the outlined strategies, actions, and the more recent reforms, starting from the full functionality of Ente Nazionale Italiano per il Turismo and the National Tourism Observatory. The most difficult and challenging phase is precisely that of implementation. Improvement in the effectiveness of tourism policies will depend, first, on concrete efforts from all the public and private entities involved and, second, on the lines and objectives set in the PST and the Piano Straordinario della Mobilità Turistica (PSMT).

Sustainability of tourism requires that it integrate the natural, cultural, and human environment of territory.

The city of Naples already has moved in this direction through its spontaneous movement of tourist development. In fact, local tourist companies have chosen a strong local connotation both from a gastronomic and cultural point of view, preferring often suppliers at km 0.

Even the human resources selected are purely local, especially in small and micro tourist enterprises. Neapolitan unemployment in 2017 is equal to 30% in the complex and reaching more than 50% in the youth band (Eurostat data), and at the same time possessing a wide range of training linked to both tourism and sustainability. For this reason, Neapolitan companies do not need to recruit human resources not resident in the territory.

Therefore, voluntary or involuntary actions aimed at developing sustainable tourism have multiplied in the area.

The actions that the Naples International Airport have chosen to carry out in this direction can be mentioned. Important actions not only for Neapolitan tourism but also for its environmental sustainability.

Always in this direction, we can mentioned the actions that have chosen to carry out the airport of Naples. The airport is an important city key not only of Neapolitan tourism but also of its environmental sustainability.

In fact, in terms of sustainability, the Naples International Airport (Gesac, 2018, p. 9) pursues the following aims:

- Promotion and enhancement of the territory as a tourist destination;
- Quality and excellence of the passengers' experience;
- Enhancement and involvement of the human resources involved in airport activities;
- Protection and respect of the natural environment;
- Community engagement;
- Contribution to the socio-economic development of the territory.

The municipality of Naples, in this direction, with order no. 48 of 22 July 2015, established an autonomous organizational unit called "Promotion and Enhancement of Sustainable Tourism, Pedestrian Walkways and City Landscape Areas".

There are other initiatives that, adopted by micro, small, and medium-sized enterprises, tend to bring Neapolitan tourism to sustainability in the long run, such as the affiliation of B&Bs to ECOBNB, the accurate segregated collection of waste imposed on tourists within the small Neapolitan hospitality industry and also on tourism businesses in general, the choice of local suppliers, and the widespread availability offered to tourists to use ecological means of transport.

5. Results and Conclusion

The case study of Naples has demonstrated how, faced with spontaneous tourism, public and private bodies and different stakeholders have hesitantly started a process of strategic planning in order to direct previous single efforts toward not only a vision of leadership in different tourist markets but also the enhancement and promotion of its unique tourist products.

With reference to the model in Figure 11.1, spontaneous tourism from the offer side can then be connected with the phenomenon of self-employment and new micro-entrepreneurship in the tourism industry. Some of these new business activities have already been designed according to sustainable thinking. For example, this is the case of B&Bs which are connected with the ECOBNB community, promoting responsible tourism with a low environmental impact.

On the other side, accommodations in Naples, especially those that are connected with a corporate brand, are oriented toward environmental sustainability and therefore employ low-impact generators to reduce CO_2 emissions and energy consumption, in addition to utilizing low-consumption LED lighting and high-efficiency heating systems.

Furthermore, these firms have focused their attention on hiring local people and on investing in training activities for young students, feeding the social pillar of sustainability. In this vein, a strong connection between universities and high schools has been established with the aim to provide new generations with job and training opportunities. Numerous local entrepreneurs, operating in the tourism industry, are closed to local association with sponsorship for social projects.

Regarding restaurants, historically, Naples' offer is shaped by local ingredients that respect the identity of the place both in terms of recipes and raw materials. Furthermore, there is a growing interest in the formula offered by the organic farmhouses that are spreading throughout the surrounding areas of Naples. These examples hint at how the offer side is organized independently and seeks to open up sustainability horizons.

As a matter of fact, this "sustainable spontaneity" prevails on the offer side since spontaneous tourism on the demand side only refers to the lack of desired push factors by the local offer, influencing the choice of travel at an induced level of marketing (before the arrival at the destination, when the tourist has to decide where to go).

The Naples case study has also shown high spontaneity on the offer side. The Naples International Airport, managed by Gesac, is benefitting from it and making numerous efforts to systematize the tourism offer, also according to sustainable principles. As the model demonstrates, at a certain point in the cycle of spontaneous tourism, actors try to create networking strategies that respond to spontaneous development. In different directions, the convention bureau (meeting industry) and the Naples International Airport (destination) are promoting the territory, moving from spontaneous tourism to "planned from spontaneity", as claimed in the model.

As shown in the model, the Naples International Airport has benefitted from the spontaneity of the city's tourist offerings as a result of its marketing activities and its nature. At the same time, this actor has also worked through specific strategies to transform spontaneous tourism flows into planned flows, attracting new airline companies, enhancing existing relationships, and promoting the destination of Naples worldwide through educational and familiarization trips.

The challenges for years to come are manifold and require a shared effort from all the actors who compose the complex picture of tourism in Italy. First, it is necessary to develop a management plan for the conspicuous tourist flows that are expected soon. This is pivotal for the handling of the risk of *overtourism* in cities and areas of major appeal; the promotion of tourism development of areas exploited below their potential; the containment of *overtourism* in the cities of greatest appeal; the enhancement and promotion of the Italian country's image in the world; the expansion of the digital services for both travelers and operators within the sector; the implementation of transversal policies that can affect tourist attractiveness; the adequate regulation of new accommodation arrangements, which should ensure a leveling of competitive conditions in relation to traditional structures, without reducing the spread of alternative forms of hospitality; the reduction of the seasonality of flows, for example, by developing congressional and fair tourism; and last, the negotiation with large global tour operators, who should work to attract growing tourist demand from emerging countries and govern its distribution on the territory over time.

To put it simply, Naples has to choose whether to be a tourist city in the strict sense of the term or a city that *also* lives on tourism. The city that lives on tourism knows how to preserve itself for its own inhabitants while selling much of itself to tourism.

Today, we are witnessing the proliferation of services for tourists and the disappearance of productive and artisanal activities and more. After many years of tourist shortages, the local tourist industry aspires to an advanced market, being, however, structured in an old-fashioned way.

The "postmodern" idea of tourism believes that cities develop it to compensate for the decline caused by deindustrialization. This is the case of England in the 1970s and 1990s, when the past became a national industry. As Robert Hewison observed in 1989, of the 1,750 museums then present in Great Britain, half had been opened after 1970: in just 16 years, as many museums had been established as in all previous centuries.

Every tourist place has a life cycle that, when it has reached its peak, inevitably tends to extinguish; each location has a receptive capacity beyond which both visitors and the local population feel discomfort in the enjoyment of the urban and natural spaces they share. The life cycle of tourist resorts can be schematically divided into three phases: discovery, maturation, and decline. Each stadium has its own receptivity threshold, so it is urgent to identify the right policy to transform Naples from a tourist city into a city that lives on tourism (Gardini, 2018).

The great risk that today the city of Naples also runs, as part of the national territory, is that the peak of interest becomes a boomerang and, therefore, instead of carrying out a marketing operation of the territory, it becomes a phenomenon of *demarketing* or *overtourism*.

In this respect, today the main problem is to adapt the city to the increasing tourist flows, implementing projects related to cleanliness, safety, and maintenance of both cultural heritage and green and marine areas.

To achieve these objectives, the systematization of efforts is a precondition for the sustainable and widespread development of a sector of paramount importance for the growth of the local economy. This is especially true when taking into account, on the one hand, the strong expansion of the international demand and, on the other, the enormous tourism potential still to be exploited. However, to make the current spontaneous tourism truly sustainable in the long term, strategic plans may not be necessary. In fact, when analyzing the territory, it is easy to grasp those strategic objectives that were already spontaneously implemented. Precise operational plans are instead needed, through the gathering of requests and needs from below and the defining of self-implementing plans to achieve the objectives that have emerged from the territory.

The spontaneous process of tourism is therefore governed today by a state of harmonious freedom which is absolutely closer to humanistic than scientific management. In fact, humanistic management aims to create a more balanced relationship between what can be exchanged on the markets and what cannot be, but which makes life interesting, that is, well-being.

Moreover, humanistic management is an operational and cognitive model that tends to transform the company into a social organization founded on the opening of the organization's borders (Minghetti, 2004), in this sense the spontaneous Neapolitan process was based on the involvement of partners and above all of customers to show, beyond a communication that has always been governed and induced, the real Naples with its strengths and not only its defects.

The idea is a step-by-step programming with specific results of limited economic value but of immediate implementation. A bottom-up participatory system could be the key to adapting the territory to the flows while improving its sustainability. The institutions, in particular the municipality of Naples, could therefore define small operational plans in accordance with the micro-entrepreneurship of local tourism, which would be aimed to solve sustainability issues within small, local areas.

Nonetheless, the city still needs macro support from the municipality and the region, especially for the communication, construction, and stabilization of its image, as well as for its cleaning and maintenance. As for communication and image building, the Campania Film Commission has a key function in this, which can be carried out through product placement policies (Sciarelli, 2014).

As for cleaning and maintenance, the municipality surely plays a decisive role. In this respect, it would be necessary to define joint operational plans focused on territorial areas, to be implemented in the short term, thus bringing immediate benefits to tourists, entrepreneurs, and citizens.

An important signal in this direction, which also confirms the humanistic managerial approach of the city, came from the municipality through the permanent forum of non-profit organizations for sustainable tourism in Naples, which, in collaboration with the University Observatory on Tourism—Federico II, organized a first training seminar in January 2020 on sustainable tourism in Naples.

The aim of the seminar was precisely to help create a common language and experiment with good practices for a vision and a unified mode of sustainable tourism offerings in the city.

In sum, tourism in the city of Naples can already be considered an industrial sector (an addition and not a replacement of manufacturing activity) in that it is able to produce value for the development of the city and its metropolitan area. This is based on the data provided by one of the most recent analytical methodologies within the sector, namely the so-called presence multiplier, which indicates how much added value activates an additional tourist presence in a specific location: at the same cost, an additional tourist presence in Italy generates as a whole 103.4 euros of added value, while in the South and in the major cities south of Rome it is close to 70 euros (Banca d'Italia, 2018).

In conclusion, it can be said that for the management of spontaneous tourism, in a city like Naples, accurate methodologies of analysis are needed and should be developed in a systemic way through the dynamics of tourist product management; the realization of a project of communication and construction of a differentiated image by tourist profile; and the implementation of individual operational plans of material and immaterial infrastructures differentiated by small areas. This could make Naples and its immense metropolitan heritage truly connected, available, and accessible to everyone, as the many examples around the world remind us.

From analysis of the case study, important reflections emerge, both upon the components of the proposed model—all are necessary at different levels—and the fact that sustainability does not enable *the tourism of spontaneity*. As a matter of fact, this phenomenon is often understood as "casualty", despite its recurrence in the country. The features highlighted concern many, if not the majority, of the Italian destinations. The national system is experiencing a positive trend that is not the result of expressed strategies; it is therefore essential to plan a sustainable development process, in order to become one of the most competitive destinations in the world, also by virtue of an efficient humanistic management model.

Our investigation is undoubtedly limited because of a focus on a specific case study, but settings and implications are highly generalizable and ascribable, with due differences, to different European contexts.

References

Assoturismo-Cst per Confesercenti. (2019). *Turismo: Assoturismo—CST, XXIII Borsa delle 100 città d'Arte*. Retrieved from www.confesercenti.it/blog/turismo-assoturismo-cst-xxiii-borsa-delle-100-citta-darte/

Banca d'Italia. (2018). *Turismo in Italia numeri e potenziale di sviluppo*. Retrieved from www.bancaditalia.it/pubblicazioni/collana-seminari-convegni/2018-0023/rapporto_turismo_finale_convegno.pdf

Becheri, E., Micera, R., & Morbillo, A. (a cura di). (2018). *XXII Rapporto sul Turismo Italiano*. CNR.

Biondi, G. (2016, aprile 21). *Il turismo a Napoli: dallo sviluppo spontaneo ad una mutazione governata, Giovedì*. Retrieved from www.nagora.org/il-turismo-a-napoli-dallo-sviluppo-spontaneo-ad-una-mutazione-governata

Boluk, K., Cavaliere, C. T., & Higgins-Desbiolles, F. (2017). Critical thinking to realize sustainability in tourism systems: Reflecting on the 2030 sustainable development goals: Guest editors. https://doi.org/10.1080/09669582.2017.1333263

Briassoulis, H. (2002). Sustainable tourism and the question of the commons. *Annals of Tourism Research*, *29*(4), 1065–1085. https://doi.org/10.1016/S0160-7383(02)00021-X

De Rosa, S., & Salvati, L. (2016). Beyond a 'side street story'? Naples from spontaneous centrality to entropic polycentricism, towards a 'crisis city'. *Cities*, *51*, 74–83. https://doi.org/10.1016/j.cities.2015.11.025

Della Corte, V. (2013). *Imprese e sistemi turistici: il management. II Edizione*. Milano: Egea.

Della Corte, V., & Sciarelli, M. (2013). Alla ricerca della sostenibilità delle destination: riflessioni e primi riscontri empirici. In M. Franch & U. Martini (Eds.), *Management per la sostenibilità dello sviluppo turistico e la competitività delle destinazioni* (pp. 325–366).

Della Corte, V., Sciarelli, F., & Del Gaudio, G. (2018). The evolution of tourism in the digital era: The case of a tourism destination. *Sinergie Italian Journal of Management*, *35*(105), 179–199.

Dolnicar, S. (2004). Insights into sustainable tourists in Austria: A data-based a priori segmentation approach. *Journal of Sustainable Tourism*, *12*(3), 209–218. https://doi.org/10.1080/09669580408667234

Dwyer, L. (2005). Relevance of triple bottom line reporting to achievement of sustainable tourism: A scoping study. *Tourism Review International*, *9*(1), 79–94. https://doi.org/10.3727/154427205774791726

Garcia, S., Luengo, J., Sáez, J. A., Lopez, V., & Herrera, F. (2012). A survey of discretization techniques: Taxonomy and empirical analysis in supervised learning. *IEEE Transactions on Knowledge and Data Engineering*, *25*(4), 734–750.

Gardini, F. (2018). *"Città turistica" o "città che vive di turismo": il dilemma non c'è, mancano le politiche, Cultura 2.0, 20 Agosto 2018, Il denaro*. Retrieved from www.ildenaro.it/citta-turistica-citta-vive-turismo-dilemma-non-ce-mancano-le-politiche/

Gesac. (2018). *Report di Bilancio di Sostenibilità, 2017*. Retrieved from www.aeroportodinapoli.it/documents/10186/34976/BdS_2018-versione-intera.pdf/fd3231a0-5f04-4831-9a2c-50800a2eb67d

Hanna, P., & Adams, M. (2019). Positive self-representations, sustainability and socially organised denial in UK tourists: Discursive barriers to a sustainable transport future. *Journal of Sustainable Tourism*, *27*(2), 189–206. doi:10.1080/09669582.2017.1358272

Heslinga, J., Groote, P., & Vanclay, F. (2019). Strengthening governance processes to improve benefit-sharing from tourism in protected areas by using stakeholder analysis. *Journal of Sustainable Tourism*, *27*(6), 773–787. doi:10.1080/09669582.2017.1408635

Hewison, R. (1989). Heritage: An interpretation. *Heritage Interpretation*, *1*, 15–23. http://statistics.unwto.org/sites/all/files/docpdf/parti.pdf

Hunter, C. (1997). Sustainable tourism as an adaptive paradigm. *Annals of Tourism Research*, *24*(4), 850–867.

ISTAT dati turismo. (2017). Retrieved from www.istat.it/it/archivio/turismo

Jovicic, D. Z. (2019). From the traditional understanding of tourism destination to the smart tourism destination. *Current Issues in Tourism*, *22*(3), 276–282. doi:10.1080/13683500.2017.1313203

Lehtonen, M. (2004). The environmental—social interface of sustainable development: Capabilities, social capital, institutions. *Ecological Economics*, *49*, 199–214. https://doi.org/10.1016/j.ecolecon.2004.03.019

Liburd, J. J., & Edwards, D. (Eds.). (2010). *Understanding the sustainable development of tourism*. Oxford: Goodfellow.

Lim, C. C., & Cooper, C. (2009). Beyond sustainability: Optimising island tourism development. *International Journal of Tourism Research*, *11*(1), 89–103. doi:10.1002/jtr.688

Martini, U. (2005). *Management e gestione delle destinazioni turistiche*. Torino, Italia: Giappichelli.

Mazzone, C. (2019). *Napoli città d'arte il turismo cresce del108,7, Ottopagine Napoli*, martedì 2 aprile 2019 alle 17.53. Retrieved from www.ottopagine.it/na/cultura/181483/napoli-citta-d-arte-il-turismo-cresce-del-108-7.shtml

Minghetti, M., & Cutrano, F. (2004). *Le nuove frontiere della cultura d'impresa*. Milano, Italia: Etas.

Paniccia, P. M. A., & Leoni, L. (2019). Co-evolution in tourism: The case of Albergo Diffuso. *Current Issues in Tourism*, *22*(10), 1216–1243. https://doi.org/10.1080/13683500.2017.1367763

Park, H. S., Rene, E. R., Choi, S. M., & Chiu, A. S. (2008). Strategies for sustainable development of industrial park in Ulsan, South Korea—From spontaneous evolution to systematic expansion of industrial symbiosis. *Journal of Environmental Management*, *87*(1), 1–13. https://doi.org/10.1016/S0160-7383(02)00025-7

Pechlaner, H., Raich, F., Beritelli, P., d'Angella, F., De Carlo, M., & Sainaghi, R. (2010). Archetypes of destination governance: A comparison of international destinations. *Tourism Review*. https://doi.org/10.1016/j.jenvman.2006.12.045

Perales, R. M. Y. (2002). Rural tourism in Spain. *Annals of Tourism Research*, *29*(4), 1101–1110.

Riege, A. M. (2003). Validity and reliability tests in case study research: A literature review with "hands-on" applications for each research phase. *Qualitative Market Research: An International Journal*, *6*, 75–86.

Ritchie, J. B., & Crouch, G. I. (2003). *The competitive destination: A sustainable tourism perspective*. CABI.

Rodriguez, I., Williams, A. M., & Hall, C. M. (2014). Tourism innovation policy: Implementation and outcomes. *Annals of Tourism Research*, *49*, 76–93. https://doi.org/10.1016/j.annals.2014.08.004

ROMA. (2018). *Turismo, Napoli sul podio: è la terza meta in Italia*, Gio 10 Mag 2018 17:00. Retrieved from www.ilroma.net/news/cronaca/turismo-napoli-sul-podio-la-terza-meta-italia

Romei, P. (Ed.). (2009). *Turismo sostenibile e sviluppo locale*. Milano, Italia: Wolters Kluwer.

Saraniemi, S., & Kylänen, M. (2011). Problematizing the concept of tourism destination: An analysis of different theoretical approaches. *Journal of Travel Research*, *50*(2), 133–143. doi:10.1177/0047287510362775

Sciarelli, F. (2014). Il Turismo Cinematografico: un inconsueto motore di sviluppo territoriale. In G. Tardivo (Ed.), *Il ruolo economico del turismo culturale nella regione transfrontaliera*. Franco Angeli Editore.

Sciarelli, S. (2007). *Il management dei sistemi turistici locali: strategie e strumenti per il governance*. Torino, Italia: Giappichelli.

Stylo. (2019, Aprile 3). *Napoli, aumenta il flusso turistico culturale ma i visitatori spendono poco, Stylo 24*. Retrieved from www.stylo24.it/economia/napoli-aumento-turismo-culturale/

Swarbrooke, J. (1999). *Sustainable tourism management*. CABI.

Tremblay, P. (2000). Sustainable tourism. *Tourism Collaboration and Partnerships: Politics, Practice and Sustainability*, 2, 314.

World Commission on Environment and Development (WCED). (1987). *Our common future*. Retrieved from http://un-documents.net/ocf-02.htm

World Tourism Organization (UNWTO). (2002). *Annual report, A year of recovery*. Retrieved from www.untwto.org

Yin, R. K. (2003). *Case study research: Design and methods*. Thousand Oaks, CA: Sage Publication.

Zhang, L., Wang, Y. P., Sun, J., & Yu, B. (2019). The sightseeing bus schedule optimization under park and ride system in tourist attractions. *Annals of Operations Research*, *273*(1–2), 587–605. https://doi.org/10.1007/s10479-016-2364-4

12 Everyday and Holiday Behaviors Regarding Sustainable Consumption Choices

Humanistic Management Perspectives for an International Destination

Mariangela Franch, Pier Luigi Novi Inverardi, Federica Buffa, and Eleonora Moratti

1. Introduction

Within the area of humanistic management studies at least two key issues can be identified:

- A humanistic management ethos includes the aim of ensuring the "dignity" of those who work within an organization; as Bowie (1999) points out, however, in management studies this aspect is often underanalyzed, or even "neglected".
- Humanistic management encourages the spread of a culture of "common goods" both within a company and among the stakeholders with whom it engages. Studies based on this assumption have usually focused on the need to adopt a truly sustainable model (thus discarding the standard economic model) and to use the common good matrix (Felber, 2015; Dyllick & Muff, 2016) to assess the extent to which stakeholders adopt humanistic management models. In tourism management studies, references to humanistic management can be traced back to the truly sustainable model within which the dimension of social sustainability encompasses attention to dignity, common goods, and other socio-cultural factors. O'Neill, Hershauer and Golden (2006, p. 34) emphasize "that the[ir] cultural characteristics play a significant role in influencing the values that sustainable enterprises would like to create". This definition provides a new insight into sustainable enterprises (SE), as suggested by Nurse (2006): SEs should incorporate culture as well as social equity, environmental responsibility, and economic viability.

In this chapter, we consider whether (and if so, how) socio-cultural factors linked to tourists' country of origin (COO) and the general awareness of environmental issues in these countries determine:

- Holiday behaviors at variance with everyday behavior;
- Different propensities to pay a premium price (levels of willingness to pay—WTP) for more sustainable products and tourism services.

With regard to the first topic, we investigate whether the culture of tourists' COO influences their vacation behaviors. This analysis enables us to assess the findings of a study by Khan and Amann (2013) which indicated that in countries with a strong culture of environmental and social concern, and where institutions and businesses are more aware of these issues, humanistic management values are more widespread and are, in turn, reflected in the behaviors (even when on holiday abroad) of the population. In other words, our research examines whether such factors as sensitivity to the importance of natural resource conservation, quality of life, the common good, and the socio-cultural elements of the community are reflected in tourist behaviors. To date, research findings in this area have been contentious. Nonetheless, any coherence in people's behaviors at home and abroad revealed by the results of our study suggests that humanistic management could support decision-makers within tourism enterprises.

With regard to the second topic, our research investigates whether tourists are willing to pay more for more sustainable tourism services, and in doing so contribute to and co-finance investments identified with environmental and cultural common goods (such as sustainable mobility, the valorization of local culture by the tourist offering, etc.).

The chapter is divided into three main sections. In the following section, we indicate the principal sources in the literature from which our three research hypotheses originated. The research methodology adopted is then described, followed by a discussion of the main findings from a study of a representative sample of 384 tourists in the area of Lake Garda (Italy). The final section presents our conclusions.

2. Theoretical Background and Research Hypotheses

2.1 COO and Sustainable Behavior

Since the publication in 1997 of Elkington's seminal book, introducing the concept of the "triple bottom line", there has been increasing interest, both in the academic community and among international organizations, in the relationship between business management and the theory and practice of sustainability (Wheeler & Elkington, 2001; Sherman, 2012).

Dyllick & Muff (2016) introduced the concept of "true business sustainability" (TBS), emphasizing that sustainability cannot only create economic benefits but also contribute to social and environmental well-being. This observation is linked with a new approach to "doing business", adopting a new business model in which the economic dimension is integrated with social and environmental goals (Boons & Lüdeke-Freund, 2013; Bocken, Short, Rana & Evans, 2014). This approach strongly emphasizes the importance of a business's capacity to look outward; the more it can do so, the more easily it will be able to implement TBS. One of the most relevant aspects consists in changing

> organizational perspectives, from an inside-out perspective with a focus on the business itself to an outside-in perspective with a focus on society and the sustainability challenges it is facing. This moves the value creation perspective from the triple bottom line to creating value for the common good.
>
> (Dyllick & Muff, 2016, p. 168)

The change in perspective required of businesses necessitates a re-evaluation of the role of the individual, the potential consumer, whose needs and consumption habits must be understood; it is also essential that steps are taken to understand the cultural background against which the process of "informing and educating customers about unsustainable choices and practices" (Dyllick & Muff, 2016, p. 167) will take place.

The connection between TBS and the principles of humanistic management emerges clearly from the foregoing, as does the importance of the cultural backgrounds of both businesses and individuals. Writing on the latter dimension, Khan and Amann (2013) placed particular emphasis on the influence that the prevailing culture in their COO can have on individuals' behavior and, more specifically, on their choices and behaviors with regard to sustainability. Since the 1990s, academic interest in this topic has been growing, as has consumer (and business) interest in environmental issues (Akehurst, Afonso & Martins Gonçalves, 2012). Roberts (1996) described and discussed ecologically conscious consumer behavior (ECCB) in the 1990s, highlighting ways in which profiles had changed since the 1960s and 1970s. A gap in the literature review (*ibidem*) emerges with regard to the influence of someone's "place of residence" on their consumer behavior. This variable has not always been taken into account; when it is, however, there is usually a positive relation between it and environmental concern. The influence of COO on ECCB continued to be an open question in subsequent studies (Straughan & Roberts, 1999) and has (at least until recently); green behaviors continued to be a debated research topic (Akehurst et al., 2012). A recent study in the field of management (Golob & Kronegger, 2019) supports Khan and Amann's findings, pointing out that COO counts, that is, sustainable behavior (or pro-environmental behavior as defined in previous studies and recalled by Golob & Kronegger, 2019) and sustainable production vary from country to country. This observation, however, while it reveals a connection between COO and pro-environmental behavior, also highlights the knowledge gap with regard to the issue, particularly in Europe, where few studies have been carried out (Gross & Telesiene, 2017), and the existing findings, moreover, are sometimes contradictory (Golob & Kronegger, 2019). The critical issues related to this research area also emerge where green behaviors are linked to other, very closely connected, dimensions, such as green attitudes. Here, too, the results are not unanimous, and the factors that support the existence of this relationship are related to specific environmental practices and/or influenced by diverse factors (Sheth, Sethia & Srinivas, 2011; Yilmazsoy, Schmidbauer & Rösch, 2015).

In the European context, Austria, Denmark, and Germany have been identified in some studies as being the most environmentally aware countries ('pro-environmentalists', 'moderate environmentalists', according to Golob & Kronegger, 2019); while Italy and the Netherlands are—to varying degrees—"moderate" and "sideline" environmentalists (*ibidem*). Similar results emerge from the Environmental Performance Index (2018) which ranks each country according to its environmental performance: Denmark, Austria, and Germany score higher on environmental protection than Italy and the Netherlands. The Special Eurobarometer 468 Report (EC, 2017) highlights the fact that awareness about environmental issues has a direct effect on inhabitants' daily life and records the differences among European countries.

García-Álvarez and Moreno (2018) proposed a composite index of environmental impact assessment for the countries of Europe (then, the EU-28) that takes into account the dimension of sustainable and efficient resource use: Sweden, Austria, Denmark, Italy, and Germany performed best on their index. Ruiz de Maya, López-López and Munuera (2011) investigated organic food consumption in a number of European countries. Ruiz de Maya et al. (2011) investigated organic food consumption in a number of European countries. In some cases (such as Denmark and Sweden), the socio-cultural dimension influences purchase intention, but in others (like Spain) this relation is not supported. The Special Eurobarometer 468 Report (EC, 2017) confirms these results: in countries such as Denmark and Sweden, the presence of the Ecolabel on a product positively influenced consumer buying. Once again, the relation between COO and pro-environmental behaviors is underlined, revealing the sensitivity in certain countries to environmentally sustainable policies and choices. Nevertheless, the need for further research to understand if and how sustainable behaviors are influenced by COO is undoubted.

2.2 COO and Sustainable Behavior on Holiday

The influence of COO on sustainable behavior has also, of course, proven of interest in tourism management studies, where there is continued disagreement about whether tourists' everyday behavior (Barr, Shaw, Coles & Prillwitz, 2010; Ganglmair-Wooliscroft & Wooliscroft, 2017) and environmental awareness (Aman Harun & Hussein, 2012; Juvan & Dolnicar, 2014) influence their sustainable behaviors while on holiday (Knezevic Cvelbar, Grün & Dolnicar, 2017). The fact that there is a gap between people's everyday and holiday green behaviors is particularly hotly debated. Anable, Lane and Kelay (2006) draw attention to this gap in their investigation of some environmentally friendly behaviors (such as energy efficiency and conservation), which reveals that people in general behave in a more 'green' way when at home than they do on holiday—an attitude which the authors describe as an attempt to be "green on balance". Barr et al. (2010) confirm this tendency: most tourists, in fact, while tending to be 'environmentally friendly', aware of the environmental impacts of their behaviors and adopting green behaviors in their everyday life, seem to use the fact that they are on holiday to "justify" fewer conscientious behaviors when away from home. Becken (2007) points out the complexity of behaving environmentally responsibly while on holiday and further emphasizes the different perceptions that people have of their responsibility to the environment in different contexts.

Ganglmair-Wooliscroft and Wooliscroft (2017), however, had different results: they draw attention to the coherence between people's everyday and holiday behaviors. This consistency is particularly marked when—whether at home or away—the activities involve similar procedures and levels of effort.

In light of the foregoing, our chapter analyses the influence of people's COO on their everyday and holiday behaviors and discusses whether their environmental awareness influences their holiday behaviors. The hypotheses that the research aims to test are therefore:

- Hypothesis 1: People's everyday sustainable behaviors are replicated while on holiday.
- Hypothesis 2: Environmental awareness influences holiday behaviors.

2.3 COO and Willingness to Pay on Holiday

Another—inescapable—topic in any analysis of tourists' sustainable behaviors is willingness to pay (WTP)—the propensity to pay a higher holiday cost in order to support sustainable investment in the destination (consistent with the truly sustainable model). The debate within tourism management studies—as in the management field generally (see Didier & Lucie, 2008)—is intense, and the findings on WTP for sustainable destinations are not unanimous, as highlighted in some studies done in maritime destinations. Cetin, Alrawadieh, Dinçer, Dincer and Ioannides (2017) investigated WTP to mitigate the negative externalities generated by tourism in Istanbul. The tourists said they were willing to pay more for their holiday if the increases were going to be invested in measures to preserve environmental and social resources. The tourists were concerned about minimizing the negative effects of tourist activities on the local environment and its inhabitants. These results are in line with those of a study by Birdir, Ünal, Birdir and Williams (2013) into tourists' WTP for the protection of three of Turkey's biggest beaches: a majority of their respondents said that they were willing to pay (up to between 1.77 and 2.33 euros a day) more for their holiday if this meant that (man-made) tidal waste was cleaned up, the beaches were enhanced, and the area's services improved. Studies by Logar and van den Bergh (2014) and Sayan, Williams, Johnson and Ünal (2011)—the former conducted to investigate tourists' WTP to prevent beach erosion in Crikvenica (Croatia), the latter examining their willingness to contribute to the protection of natural resources in the coastal zone of Antalya (Turkey)—also provide convincing evidences of tourists' propensity to contribute financially to more sustainable resource management.

Two other studies, however, demonstrate less environmental concern on the part of tourists (evidenced by a lower WTP): Pulido-Fernández and López-Sánchez (2016), conducted on the Western Costa del Sol (Andalusia, Spain), and Blakemore, Williams, Coman, Micallef and Unal (2002), conducted in seaside resorts in Malta, Romania and Turkey.

Our research, as part of this debate on tourists' WTP to support sustainable practices and initiatives, includes a further hypothesis:

- Hypothesis 3: Tourists are willing to pay for more sustainable tourism services.

Based on this evidence of tourists' self-declared WTP, we can point to demand-side propensities and requirements which it would be possible for policy-makers to exploit in order to direct investment toward the economic, environmental, and socio-cultural development of the territory, consistent with the principles of humanistic management.

3. Methodology

3.1 Sample and Data Analysis

The area of investigation was Lake Garda, a well-known destination in northern Italy, chosen because it provided an opportunity to test our research hypotheses in an international tourist destination which suffers, particularly in the summer, from the effects of high tourist flows. The sustainable management of tourist flows is thus a challenge faced both by destination managers and by public actors endeavoring to guarantee the local community's quality of life.

A face-to-face questionnaire was administered to a statistically significant sample of 384 tourists (from Italy, Austria, Germany, Denmark, and the Netherlands) to investigate the existence of a link between green-oriented behaviors at home and on holiday. The stratification of the sample according to COO was weighted according to the relative share of tourist presences (in the Lake Garda area) from the five countries included in the survey, for the years 2015–16.

A payment card was attached to the questionnaire to collect the interviewees' statements about their WTP. There were Italian, English, and German versions of the questionnaire, which was administered following a sequential sampling scheme in the main tourist areas of the lake during July and August 2017.

A total of 276 people from the aforementioned countries responded to the questionnaire. Different hypotheses of interest were tested, with the aim of ascertaining whether or not certain behaviors adopted by tourists at home and on holiday had a significant correlation with their COO. In doing this, it was necessary to take the following factor into account: the statistical methods principally employed in the analysis of collected data use the information contained within the combined distribution (contingency tables) to investigate any significant association between the various elements under investigation (behaviors, WTP, levels of awareness, and/or knowledge, etc.) and between these elements and a respondent's COO. Because these analysis methods are sensitive to the presence of low-frequency cells in the tables, a reaggregation of the tourists into three (geographically and/or culturally, see section 4.1) similar groups was sometimes undertaken, exploiting the additivity of the chi-squared statistic:

- Group 1: Italian tourists ($n_1 = 53$)
- Group 2: Austrian and German tourists ($n_2 = 169$)
- Group 3: Dutch and Danish tourists ($n_3 = 54$)

This reaggregation is also justified in relation to the environmental awareness declared by the tourists involved in the survey. On a 1-to-10 scale, environmental awareness associated with the tourists from each country included in the analysis is as follows: Austria (8.36), Germany (7.55), Italy (6.50), Denmark (6.64), and the Netherlands (6.00).

The gender composition of the sample reveals a majority of Italian (64.15%), German (62.67%), and Danish (57.14%) women, and Dutch (55.00%) and Austrian (57.89%) men. The 36–55 age group was the largest (56.16% of respondents). Across nationalities, the best-represented education level was the high school diploma (Italy, 54.72%). Among the Dutch and Danish tourists, graduates (with bachelor's degrees) comprised 37.50% and 35.71%, respectively of the respondents, while a significant proportion of the Austrians and Germans possessed a lower educational level (Austria, 36.84%; Germany, 30.67%).

As a general rule, in the comparison of everyday versus holiday behaviors, the hypothesis being tested—which is indicated by H_0 (or null hypothesis)—is that of independence, understood as nonconformity between everyday and holiday behaviors (for details on the analysis of contingency tables, see Agresti, 1992, 2018). The *p-value* of each hypothesis, representing the probability of hypothesis H_0 in the light of the information contained in the data, is given. A low p-value signifies the rejection of the null hypothesis in favor of the alternative hypothesis which usually represents the research hypothesis and which, as such, challenges the status quo established by the null hypothesis.

For example:

> H_0: Everyday and holiday behaviors differ with regard to waste separation versus H_1: There is NO difference between everyday and holiday behaviors with regard to waste separation.
>
> (p-value = 0.00023)

This result leads to the rejection of H_0 in favor of H_1 since the low p-value indicates insufficient evidence for H_0.

To further confirm the decision in favor of H_0 and to have a reference by which to establish when the measured p-value is low, it is usual to compare the p-value with a significance level α (here defined as 0.05) which thus serves as the base value and allows us to reject H_0 whenever the p-value is less than α.

3.2 *Testing the Research Hypothesis*

To investigate whether people's everyday sustainable behaviors are replicated while on holiday (hypothesis 1), it was important to measure behaviors related to waste separation (item a), the use of public transport (item b), and how often people bought local products (item c). The decision to analyze these items was determined by their importance for describing pro-environmental behavior, as demonstrated by Golob and Kronegger (2019) and the aforementioned European study (EC, 2017). These behaviors contribute to the creation of a common good and produce socio-economic benefits for the territory and the local community.

Tourists' choices to separate waste (item a) and use public transport (item b) enable us to assess the extent to which they can contribute to reducing the negative environmental impact of their holiday presence. The careful separation of waste and greater use of public transport naturally help to reduce the anthropic impact on, and negative externalities for, the environment, thereby making the destination more sustainable.

The measurement of how often people bought local products (item c), on the other hand, allows us to assess their contribution to economic sustainability and the strengthening of short supply chains.

The intensity of the behaviors related to the three aforementioned items, both in everyday life and on holiday, was measured using a five-level Likert scale. Behavior coherence in the two different situations was studied by means of the correspondence analysis technique.

To verify whether environmental awareness has a significant influence on holiday behaviors (hypothesis 2), each interviewee was asked to evaluate their own level of information on environmental issues by using a traditional score scale from 1 (min) to 10 (max).

Starting from the assumption that environmental awareness—as well as knowledge—is the self-awareness of the possession of information on environmental issues (see Golob & Kronegger, 2019) on the basis of the level of information, each of the five respondent groups was divided into two subgroups:

- The first—made up of those who were on average more informed—was simply obtained by collecting the individuals with above-average (as a group) environmental awareness;
- The second—made up of those who were on average less informed—was obtained by collecting the individuals with below-average (as a group) environmental awareness.

The next step was to verify the hypothesis relative to the effect of levels of awareness on the three identified items (a, b, c), using a box plot comparative analysis of the subgroups of the members of the above-average and below-average subgroups (see Agresti (2018) for more details on the statistical techniques adopted). We chose to perform the different subgroup comparisons using the box plots; they represent a particularly appropriate tool since they include information on different—but complementary—aspects such as the center, dispersion shape, and concentration of the dataset upon which they are built; exploiting these characteristics we were able to capture the most relevant part of the information contained in the data, which facilitated the identification of any similarities and dissimilarities among the subgroups.

The box plot comparative analysis allows us to examine the effects of the (various) levels of environmental awareness on the (various) behaviors adopted by the two subgroups in the three areas of investigation (a, b, c). In this part of the analysis, groups 2 and 3 were further subdivided in relation to the geographical origin of the components—German and Austrian, Dutch and Danish, respectively—with the aim of verifying whether geographical origin and information level could together help to explain the dynamics of green-oriented holiday behaviors.

The third hypothesis regards WTP: Tourists are willing to pay for more sustainable tourism services. Its aim is to investigate whether the premium price can support investments that benefit the territory and the local community within a humanistic management framework. The payment card was chosen as the measurement tool in this investigation of WTP; it consists of a survey form where prices increase from 0.10 euro/day up to 0.50 euro/day (or more). The respondents had to choose the option which represented the maximum that they would be willing to pay.

The areas in which they were asked to express their WTP were:

1) Increasing the efficiency of waste disposal systems
2) Improving public transport
3) Increasing the availability of local products
4) Increasing the destination's renewable energy use
5) Increasing funding for the local population and culture
6) Increasing involvement of the local community in tourist offerings
7) Preserving better the authenticity of the area
8) Purchasing of organic products
9) Purchasing of green/eco-friendly services

Interviewees were provided with some extra information about items 3 and 7. With regard to "local products", it was decided not to give a precise kilometer range, since "local" is a subjective term and closely linked to each individual's knowledge of the territory: a local product is thus one which the tourist her/himself considers to be local. The "authenticity of the area" is understood as the unique features of the territory. It was suggested to the interviewees that the territory's authenticity lies in its most distinctive feature: the natural beauty of Lake Garda. Therefore, the aim was to understand tourists' WTP to preserve the landscape of the area.

To assess the potential economic benefits for the destination if one or more of the aforementioned policies were implemented, their economic impact was estimated multiplying the average daily premium price per person by the number of nights spent by tourists from the five countries (data from 2015–2016). It is widely acknowledged that respondents' statements and the ways in which they actually choose to allocate their money almost always differ significantly: the so-called social desirability bias (Grimm, 2010). This mechanism, which distorts responses, is well known in the psychology and social science literature; it pushes people to provide responses generally considered more—morally and socially—acceptable, coherent with the (perceived) norm and the social context of reference. To cope with this response distortion, which can produce an unrealistic overestimation of the interviewees' effective willingness to pay, a corrective mechanism for the WTP values was adopted, which applied the highest value (equal to 3) of a correction factor interval given by [2.6, 3] to the WTP values as suggested by Loomis (2011); consequently we obtaining the so-called bias-corrected WTP.

4. Results

4.1 People's Everyday and Holiday Sustainable Behaviors

Our research examined whether (and what) links existed between people's everyday and holiday behaviors, identifying the items that had most influenced any behavior gaps. This analysis enabled us to answer the first research question: COO influences tourists' holiday behaviors. On the other hand, there seems to be no significant association between the accommodation type and the holiday behaviors of tourists: environmental attitudes are uniformly distributed across the different types of accommodation.

With regard to waste separation, the results are statistically significant for all three groups (Table 12.1).

Italian tourists tend not to modify their everyday behavior while on holiday: the data association is very robust, and the Italians are consistently conscientious in their separation of waste. Tourists from the Netherlands and Denmark follow a similar (although somewhat less rigorous) pattern. The values on the main diagonal of the contingency table are therefore more closely associated, indicating strong uniformity between everyday and holiday behaviors. In contrast, the segment that includes the Austrians and the Germans reveals a weaker association between virtuous behaviors in everyday life and on holiday. Among the respondents in this segment, 61% said that they were less attentive in disposing of their waste while on holiday, compared with a higher percentage registered at home, where 91% of the respondents carefully separates their waste.

With regard to the frequency with which they used public transport (Table 12.2), two of the three groups expressed low approval of the routes in the destination generally. For groups 1 and 3 in particular, there is not enough evidence in the data to support the hypothesis of coherence between everyday and holiday behavior. The group comprising the

Table 12.1 People's everyday sustainable behaviors are replicated while on holiday (hypothesis 1—item a)

Behavior	*Confirmation of the first research hypothesis*
Waste separation (item a)	Group 1: YES (p-value < 0.0001) Group 2: YES (p-value $= 0.026$) Group 3: YES (p-value $= 0.002$)

Source: Authors' own elaboration

Table 12.2 People's everyday sustainable behaviors are replicated while on holiday (hypothesis 1—item b)

Behavior	*Confirmation of the first research hypothesis*
Use of public transport (item b)	Group 1: NO (p-value $= 0.428$) Group 2: YES (p-value < 0.0001) Group 3: NO (p-value $= 0.088$)

Source: Authors' own elaboration

Table 12.3 People's everyday sustainable behaviors are replicated while on holiday (hypothesis 1—item c)

Behavior	*Confirmation of the first research hypothesis*
Frequency of purchase of local products (item c)	Group 1: NO (p-value = 0.117) Group 2: YES (p-value = 0.000) Group 3: YES (p-value = 0.008)

Source: Authors' own elaboration

Table 12.4 People's everyday sustainable behaviors are replicated while on holiday (hypothesis 1—items a, b, c). Analysis by tourists' COO.

COO	*Group*	*Item a*	*Item b*	*Item c*
Italy	1	YES	NO	NO
Austria and Germany	2	YES	YES	YES
Denmark and the Netherlands	3	YES	NO	YES

Source: Authors' own elaboration

Austrian and German tourists, on the other hand, did register statistically significant results, from which we can conclude that people's limited use of public transport at home is closely linked with the same behavior while on holiday.

The respondents' purchase of local products (Table 12.3) in groups 2 and 3 did not differ significantly whether they were at home or on holiday. The data on the Italian tourists, in particular, show lack of homogeneity in behaviors (p-value = 0.117), while the results for the second and third groups highlight a decidedly greater and more significant coherence of local product purchasing behavior at home and on holiday (p-value Group 2 = 0.000; p-value Group 3 = 0.008).

Table 12.4 summarizes the main findings just described, subdivided according to COO.

4.2 Awareness of Environmental Issues and Holiday Behaviors

Analyzing holiday behaviors in greater depth, we considered whether the level of environmental awareness declared by the respondents and the need they felt to behave in more sustainable ways while on holiday was a variable which could potentially explain behaviors. (Here, it is useful to remember that, with regard to levels of awareness, each group was divided into two subgroups: that of the more aware than average tourist, and that of the less aware than average.)

The analysis of waste separation—carried out by comparing the box plots corresponding to the two subgroups—revealed that among the

Italian tourists there were significant differences between the two aforementioned subgroups. Although almost all of the respondents demonstrated virtuous behaviors independently of their declared level of awareness, those who said that they were less aware in fact demonstrated behaviors that were more careful and more homogeneous with regard to environmental protection. Most of the respondents, whichever subgroup they were in, fell clearly within the "extremely conscientious" behavior segment with regard to waste separation.

The Austrian tourists also demonstrated different behaviors depending on which subgroup they were in. Although the average and median values for the subgroups did not reveal substantial differences, the holiday behaviors of those who declared themselves to be below average in their level of environmental awareness tended to differ more. The subgroup of above-average tourists revealed noticeably more homogeneous behaviors—further proof of the impact of higher awareness levels. In the German group, too, the average and median values did not seem to differ greatly between the two subgroups, although the more aware individuals evidenced more homogeneous behavior because of higher awareness levels. Similar results were found for the Danish group.

The virtuous holiday behaviors of the Dutch tourists, however, did not seem to be linked to greater awareness. In fact, just as for the Italians, it seems that the subgroup of less aware Dutch tourists actually reveals itself to be (very slightly) more conscientious than average in its waste separation.

Moving now to the frequency of public transport use, we see that no significant difference emerges in the behaviors of Italian tourists in the two subgroups.

With regard to groups 2 and 3 we can see that:

- For the Austrian tourists, although in general their use of public transport was relatively low, the hypothesis that awareness levels and behavior choices are associated is confirmed. In fact, 50% of the more aware individuals are above the maximum value in the subgroup of less aware tourists.
- The German tourists partially upset the results—their relative levels of awareness did not appear to influence use of sustainable mobility while on holiday.
- The Dutch and Danish tourists confirm the hypothesis that awareness levels have a positive influence on behaviors, although for these two groups the influence of awareness was weak.

Analyzing the box plot of the frequency with which local products were purchased, we find that the less aware Italian and German tourists did not buy these products significantly less frequently than their more aware compatriots. The hypothesis that greater environmental awareness

Table 12.5 Environmental awareness influences holiday behaviors (hypothesis 2)

COO	*Waste separation (item a)*	*Use of public transport (item b)*	*Frequency of purchase of local products (item c)*
Italy	NO	NO	NO
Austria	YES	YES	YES
Germany	YES	NO	NO
Denmark	YES	YES, weak	YES
Netherlands	NO	YES, weak	YES

Source: Authors' own elaboration

leads to more frequent purchases of local products is therefore not confirmed (for these two groups) by the data.

The data on the purchasing habits of the Austrian tourists, in contrast, do confirm this hypothesis: most of the more aware individuals bought local products more frequently than did the members of the less aware subgroup.

The more environmentally aware Danish tourists revealed a clear propensity to purchase local products and were noticeably more likely to do so than the subgroup of less aware tourists.

The Dutch tourists also confirmed the hypothesis that frequency of purchase is associated with higher levels of awareness.

Table 12.5 summarizes the main findings described in this section and subdivided by COO.

4.3 Willingness to Pay for More Sustainable Tourism Services

To assess the importance that tourists accord to sustainability, we measured their WTP for certain environmentally and socially responsible choices made by operators in the destination.

Regardless of where they came from, 88% of the respondents said they were willing to pay a premium price of between 0 and 9.99% more than their usual daily holiday expenditure to stay in a more sustainable destination. About 10% said they would be willing to pay between 10 and 19.99% more than their usual daily spend.

In general, we can see that the lowest WTP is associated with potential improvements to public transport and the higher value with greater attention to the preservation of place authenticity (Table 12.6); each value reflects the percentage of additional WTP that tourists wished to allocate to a given investment in sustainability. The main differences among the countries can be summarized as follows:

- Italian tourists' WTP is principally oriented toward the preservation of place authenticity (13.59%), the increased use of renewable

Table 12.6 Tourists are willing to pay for more sustainable tourism services (hypothesis 3)

Potential sustainable action	*Total* (*n* = 276)	COO				
		Italy	*Austria*	*Germany*	*Denmark*	*Netherlands*
More efficient waste disposal	11.28%	12.44%	13.49%	10.34%	11.26%	11.43%
Better public transport	9.14%	9.21%	7.21%	9.07%	9.32%	8.65%
Increased use of renewable energy in the destination	12.00%	12.74%	11.83%	10.94%	10.68%	12.82%
Greater availability of local products	11.29%	9.89%	13.49%	12.13%	11.84%	10.9%
More support for local culture and population	11.61%	10.49%	8.87%	11.76%	12.23%	13.46%
More integration of local community in the tourism offering	10.58%	10.92%	8.32%	10.73%	11.26%	9.19%
Greater emphasis on preservation of place authenticity	12.77%	13.59%	14.23%	12.9%	10.49%	12.07%
Purchase of organic products	10.34%	10.35%	11.65%	10.94%	10.87%	8.87%
Purchase of green/eco-friendly services	10.99%	10.38%	10.91%	11.2%	12.04%	12.61%

Source: Authors' own elaboration

energy in the territory (12.74%), and more efficient waste disposal (12.44%). This last result is consistent with their greater conscientiousness with regard to waste separation.

- The Austrian tourists declared a willingness to pay a premium of more than 14% for the “preservation of place authenticity”. The WTP values for a more efficient waste disposal system and for greater availability of local products are both greater than 13%.
- The German tourists, too, showed a clear preference for the protection of authenticity and wanted to allocate 12.9% of the total WTP to this item, followed by availability of local products (12.13%) and support for population and culture (11.76%).
- Tourists from the Netherlands are particularly interested in the culture and local population and declared themselves willing to allocate 13.46% of the total price increase to their preservation. Their WTP for increased renewable energy use was also significant (12.82%).
- The Danish tourists demonstrated themselves to be most willing to pay for activities and initiatives that supported the local population and culture (12.23%). Next came the greater availability of green services (12.04%) and local products (11.84%) in the destination.

Overall, the tourists declared themselves to be willing to increase their average daily spend per person by about 3 euros/day (Italian, 3.16 euros; Austrian, 2.84 euros; German, 3.56 euros; Dutch, 2.34 euros, Danish, 3.68 euros) if this money was invested in the destination’s sustainability.

An interesting relationship between WTP and accommodation type was also observed: the subjects who demonstrated the highest WTP (40–50 cents per day) were those who stayed in tourism villages, private house rentals, or B&Bs. These results were valid for all the investigated dimensions except organic products and the preservation of place authenticity.

Another aspect taken into account when analyzing the data, in particular the information contained in the open questions, relates to the question of whether a tourist’s WTP can be linked to their expectation of benefits connected to the services for which their WTP is being investigated. Our findings indicate that:

- Respondents who said that they were satisfied with a particular service declared a high (in 41% of cases), medium (also 41%), and low (18%) WTP for improvement to that service;
- Respondents who said that they were not satisfied with a particular service declared a high (44%), medium (39%), and low (17%) WTP for improvement to that service.

The WTP in relation to the expected benefits from a particular service presents—initially—a similar trend, both when people are satisfied and when they are not. Further stressing the analysis of the data, when people

Table 12.7 Stated and bias-corrected WTP

Group	*COO*	*Stated WTP*	*Bias-corrected WTP*
1	Italy	12,524,094 €	4,174,698 €
2	Austria	2,672,379 €	890,793 €
	Germany	36,081,565 €	12,027,188 €
3	Denmark	4,301,513 €	1,433,837 €
	Netherlands	4,193,040 €	1,397,680 €
	Total	**59,772,591 €**	**19,924,196 €**

Source: Authors' own elaboration

are not satisfied, we can link the differences in the WTP distribution to their expectation of greater benefits as a result of improvements to a service. On the contrary, when people are satisfied, the WTP is associated with the strength of their satisfaction.

The estimate of total WTP, reconstructed on the basis of the real nights measured for the five tourist nationalities chosen for our survey, is almost 60 million euros. The breakdown per group/country is described in Table 12.7. To avoid excessive distortion related to social desirability bias, a prudential calibration of WTP by a factor of three was used (see section 3). The resulting total economic value is therefore under 20 million euros, to which each group/country contributes as indicated in the column "bias corrected" WTP.

These extra resources could be directed by public decision-makers to the areas that Lake Garda's tourists, according to these findings, appear to prioritize: increased preservation of place authenticity, initiatives that support the local population and culture, waste disposal, and greater use of renewable energy.

5. Conclusions

Starting with an investigation of tourist behaviors, this work then discusses how humanistic management can inspire sustainable tourism policies.

The sustainable approach, implemented on the basis of its three pillars—social, environmental, and economic—seemed to be the model best suited to our purposes. Environmental and social equity are both expressions of the culture of a country and can be translated into behaviors attentive to the latter's protection and conservation.

Taking this as our basic premise, we conducted a study on the cultural profiles of tourists, framed in terms of the coherence between their sustainable behaviors and environmental awareness on holiday and in their COO. The propensity of these tourists to pay a premium price for improvement of the destination's sustainability was also assessed.

Some of our most significant data revealed the popularity of investment in the preservation of the landscape, the local culture and population, and greater utilization of low-impact energy sources.

Business leaders and policy-makers who want to work in accordance with the principles of humanistic management will find much of value in the findings of this study. It would be particularly advantageous for them to:

- Acquaint themselves with the behavior profiles of the tourists who stay in the destination and understand the extent to which these reflect the cultures of their COO. When the two profiles are coherent, they can serve as useful guides in local policy-making;
- Identify investment opportunities on the grounds of tourists' willingness to pay a higher price for their stay. Such choices could strengthen the inter- and intra-generational pact implicit in both sustainable development approaches and humanistic management.

In relation to the first point, the answers to two questions seem significant: if and how the culture of a tourist's COO is reflected in their holiday behaviors and if and how their level of awareness can predict environmentally friendly holiday behavior.

Khan and Amann (2013)'s hypothesis that the culture of a tourist's COO is reflected in their holiday behaviors—a topic of considerable debate in the tourism management literature—is partially confirmed by our findings. In fact, the everyday and holiday behaviors of the German and Austrian tourists tend (to varying degrees) to be coherent. Dutch and Danish tourists maintain their virtuous everyday behaviors while on holiday for waste separation and purchasing of local products, while Italian tourists tend to be coherent only for waste separation. The attempt to be "green on balance" described in Anable et al. (2006), Barr et al. (2010) and Becken (2007) in terms of the three observed behaviors, can be seen—to varying degrees—in all the groups of tourists and, although they all tend to behave consistently in both contexts, greater environmental concern is evident in their COO.

Our findings confirm that the Danish and Austrian tourists' awareness of environmental issues is associated with a stronger environmental ethos. These results confirm those of Golob and Kronegger (2019) and the findings of the Environmental Performance Index (2018). The Dutch tourists' environmental awareness could predict their propensity to use public transport and buy local products. In contrast, the German tourists' environmental awareness was found only to influence their propensity to separate waste. Finally, no difference was found between the more and the less aware Italian (sub)groups. These results partially confirm the aforementioned studies.

With regard to the second point (WTP), the results show that, regardless of their provenance, most of the tourists were willing to pay a premium price of between 0 and 9.99% of their average holiday cost, provided that the increased revenue was used by policy-makers to adopt humanistic management strategies aimed at making the destination more sustainable. Specifically, the most popular areas of future investment were the preservation of place authenticity and the increased valorization of local culture, while the lowest values were associated with improving public transport. While these elements are the preferred recipients of the extra money tourists are willing to pay at the aggregate level, the five countries differ in their item preferences. The preservation of place authenticity and increased use of renewable energy are particularly popular among Italian and Austrian tourists. The Germans, Dutch, and Danish, on the other hand, are willing to pay more to support local culture and population. COO can thus be seen not only to influence behaviors but also the allocation of tourists' WTP for more sustainable tourist services.

The overall WTP—even the conservative figure decided upon here—could fund significant investment in the local territory and community, thereby fully aligning the choices of policy-makers and operators with key humanistic management objectives.

References

Agresti, A. (1992). A survey of exact inference for contingency tables. *Statistical Sciences*, *7*(1), 131–177. doi:10.1214/ss/1177011462

Agresti, A. (2018). *Statistical methods for the social sciences* (5th ed.). Harlow, Essex: Pearson.

Akehurst, G., Afonso, C., & Martins Gonçalves, H. (2012), Re-examining green purchase behaviour and the green consumer profile: New evidences. *Management Decision*, *50*(5), 972–988. doi:10.1108/00251741211227726

Aman, A., Harun, A., & Hussein, Z. (2012). The influence of environmental knowledge and concern on green purchase intention the role of attitude as a mediating variable. *British Journal of Arts and Social Sciences*, *7*(2), 145–167.

Anable, J., Lane, B., & Kelay, T. (2006). *An evidence base review of public attitudes to climate change and transport behaviour*. London: Department of Transport.

Barr, S., Shaw, G., Coles, T., & Prillwitz, J. (2010). 'A holiday is a holiday': Practicing sustainability, home and away. *Journal of Transport Geography*, *18*(3), 474–481. doi:10.1016/j.jtrangeo.2009.08.007

Becken, S. (2007). Tourists' perception of international air travel's impact on the global climate and potential climate change policies. *Journal of Sustainable Tourism*, *15*(4), 351–368. doi:10.2167/jost710.0

Birdir, S., Ünal, Ö., Birdir, K., & Williams, A. T. (2013). Willingness to pay as an economic instrument for coastal tourism management: Cases from Mersin, Turkey. *Tourism Management*, *36*, 279–283. doi:10.1016/j.tourman.2012.10.020

Blakemore, F. B., Williams, A. T., Coman, C., Micallef, A., & Unal, O. (2002). A comparison of tourist evaluation of beaches in Malta, Romania and Turkey. *World Leisure Journal*, *44*(2), 29–41. doi:10.1080/04419057.2002.9674268

Bocken, N. M. P., Short, S. W., Rana, P., & Evans, S. (2014). A literature and practice review to develop sustainable business model archetypes. *Journal of Cleaner Production*, *65*, 42–56. doi:10.1016/j.jclepro.2013.11.039

Boons, F., & Lüdeke-Freund, F. (2013). Business models for sustainable innovation. *Journal of Cleaner Production*, *45*, 9–19. doi:10.1016/j.jclepro.2012.07.007
Bowie, N. E. (1999). *Business ethics: A Kantian perspective*. Oxford: Blackwell.
Cetin, G., Alrawadieh, Z., Dinçer, M. Z., Dincer, F. I., & Ioannides, D. (2017). Willingness to pay for tourist tax in destinations: Empirical evidence from Istanbul. *Economies*, *5*(2), 21. doi:10.3390/economies5020021
Didier, T., & Lucie, S. (2008). Measuring consumer's willingness to pay for organic and fair trade products. *International Journal of Consumer Studies*, *32*, 479–490. doi:10.1111/j.1470-6431.2008.00714.x
Dyllick, T., & Muff, K. (2016). Clarifying the meaning of sustainable business: Introducing a typology from business-as-usual to true business sustainability. *Organization & Environment*, *29*(2), 156–174. doi:10.1177/1086026615575176
EC—European Commission. (2017). *Attitudes of European citizens towards the environment (Special Eurobarometer 468)*. Retrieved from https://data.europa.eu/euodp/en/data/dataset/S2156_88_1_468_ENG
Elkington, J. (1997). *Cannibals with forks: The triple bottom-line of 21st century business*. Oxford: Capstone.
Environmental Performance Index. (2018). *2018 EPI results*. Retrieved from https://epi.envirocenter.yale.edu/epi-topline?country=&order=field_epi_rank_new&sort=asc
Felber, C. (2015). *Change everything: Creating an economy for the common good*. Wien: Zed Books.
Ganglmair-Wooliscroft, A., & Wooliscroft, B. (2017). Ethical behaviour on holiday and at home: Combining behaviour in two contexts. *Journal of Sustainable Tourism*, *25*(4), 589–604. doi:10.1080/09669582.2016.1260573
García-Álvarez, M. T., & Moreno, B. (2018). Environmental performance assessment in the EU: A challenge for the sustainability. *Journal of Cleaner Production*, *205*, 266–280. doi:10.1016/j.jclepro.2018.08.284
Golob, U., & Kronegger, L. (2019). Environmental consciousness of European consumers: A segmentation-based study. *Journal of Cleaner Production*, *221*, 1–9. doi:10.1016/j.jclepro.2019.02.197
Grimm, P. (2010). *Social desirability bias. Wiley international encyclopedia of marketing* (Vol. II). Hoboken, NJ: John Wiley & Sons, Ltd.
Gross, M., & Telesiene, A. (2017). Introduction: How green are Europeans? In A. Telesiene & M. Gross (Eds.), *Green European: Environmental behaviour and attitudes in Europe in a historical and cross-cultural comparative perspective* (pp. 1–10). Oxon: Routledge.
Juvan, E., & Dolnicar, S. (2014). The attitude-behaviour gap in sustainable tourism. *Annals of Tourism Research*, *48*, 76–95. doi:10.1016/j.annals.2014.05.012
Khan, S., & Amann, W. (Eds.). (2013). *World humanism*. Houndmills: Palgrave Macmillan.
Knezevic Cvelbar, L., Grün, B., & Dolnicar, S. (2017). Which hotel guest segments reuse towels? Selling sustainable tourism services through target marketing. *Journal of Sustainable Tourism*, *25*(7), 921–934. doi:10.1080/09669582.2016.1206553
Logar, I., & van den Bergh, J. C. J. M. (2014). Economic valuation of preventing beach erosion: Comparing existing and non-existing beach markets with stated and revealed preferences. *Journal of Environmental Economics and Policy*, *3*(1), 46–66. doi:10.1080/21606544.2013.863742

Loomis, J. (2011). What's to know about hypothetical bias in stated preference valuation studies? *Journal of Economic Surveys*, *25*(2), 363–370. doi:10.1111/j.1467-6419.2010.00675.x

Nurse, K. (2006). *Culture as the fourth pillar of sustainable development. Small states: Economic review and basic statistics* (Vol. 11). London: Commonwealth Secretariat.

O'Neill, G. D., Jr., Hershauer, J. C., & Golden, J. S. (2006). The cultural context of sustainability entrepreneurship. *Green Management International*, *55*, 33–46.

Pulido-Fernández, J., & López-Sánchez, Y. (2016). Are tourists really willing to pay more for sustainable destinations? *Sustainability*, *8*(12), 1240. doi:10.3390/su8121240

Roberts, J. A. (1996). Green consumers in the 1990s: Profile and implications for advertising. *Journal of Business Research*, *36*(3), 217–231. doi:10.1016/0148-2963(95)00150-6

Ruiz de Maya, S., López-López, I., & Munuera, J. L. (2011). Organic food consumption in Europe: International segmentation based on value system differences. *Ecological Economics*, *70*(10), 1767–1775. doi:10.1016/j.ecolecon.2011.04.019

Sayan, S., Williams, A. T., Johnson, D. E., & Ünal, Ö. (2011). A pilot study for sustainable tourism in the coastal zone of Antalya, Turkey: Tourists, turtles or both? *Journal of Coastal Research*, *64*, 1806–1810.

Sherman, W. R. (2012). The triple bottom line: The reporting of doing well & doing good. *Journal of Applied Business Research*, *28*(4), 673–682. doi:10.19030/jabr.v28i4.7051

Sheth, J. N., Sethia, N. K., & Srinivas, S. (2011). Mindful consumption: A customer-centric approach to sustainability. *Journal of the Academy of Marketing Science*, *39*(1), 21–39. doi:10.1007/s11747-010-0216-3

Straughan, R., & Roberts, J. (1999). Environmental segmentation alternatives: A look at green consumer behavior in the new millennium. *Journal of Consumer Marketing*, *16*(6), 558–575. doi:10.1108/07363769910297506

Wheeler, D., & Elkington, J. (2001). The end of the corporate environmental report? Or the advent of cybernetic sustainability reporting and communication. *Business Strategy and the Environment*, *10*(1), 1–14. doi:10.1002/1099-0836(200101/02)10:1<1::AID-BSE274>3.0.CO;2-0

Yilmazsoy, B., Schmidbauer, H., & Rösch, A. (2015). Green segmentation: A cross-national study, *Marketing Intelligence & Planning*, *33*(7), 981–1003. doi:10.1108/MIP-12-2013-0201

13 Memorable Experiences in Slow Tourism

An Empirical Investigation of Camping

Alessandro M. Peluso, Virginia Barbarossa, Verdiana Chieffi, and Gianluigi Guido

1. Introduction

Camping is a form of tourism in which tourists stay in strict connection with nature. In Europe, 12% of all nights spent in tourist accommodations in 2017 were spent at campsites (Eurostat, 2019). Camping entails a wide set of activities that are typically performed at a slow pace (Brooker & Joppe, 2013), such as sport activities, hiking, socializing with other tourists and local people, and staying in strict contact with the local culture (Mikulić, Prebežac, Šerić & Krešić, 2017; Moira, Mylonopoulos & Kondoudaki, 2017). As such, camping falls within the conceptual domain of *slow tourism* (Blichfeldt & Mikkelsen, 2015), which is a contemporary phenomenon characterized by tourists' tendency to travel, stay in a destination, and visit places at a decelerated pace (Dickinson & Lumsdon, 2010). Slow tourism is the opposite of traveling with the fastest transportations available, taking a quick look at tourist attractions, and staying in a destination for very short periods. In contrast, slow tourists love to move from one place to another slowly (e.g., by walking or bicycling), remain in a destination for long enough to discover its unique aspects, and immerse themselves in local traditions and culture—all in an effort to enrich themselves as human beings (Oh, Assaf & Baloglu, 2016). Given this desire to *inhabit* places rather than merely *visit* them, it is feasible to consider camping, and the broader concept of slow tourism, as humanistic.

This chapter presents an empirical investigation of camping, with the aim of bolstering the body of studies that adopt a humanistic perspective on tourism. Derived from humanist philosophy (Caton, 2016), such a perspective has emerged in business management (Melé, 2003) and the tourism industry (Botterill, 1989) in order to postulate that human development should be the ultimate goal of economic activities. In tourism, a humanistic perspective essentially means that the primary goal of tourist studies should be understanding how tourism can contribute to the satisfaction of human needs, the development of human values, and the achievement of happiness and well-being. Consistent with this view, the empirical study presented herein served two objectives. One was to assess the dimensions of camping that could arouse positive feelings (e.g., enjoyment, satisfaction), stimulate human intellect and culture, and ultimately transform tourists' holidays at campsites into unforgettable experiences that contribute to campers' human development. The other objective was to identify those experiential dimensions that might increase campers' positive responses in terms of satisfaction, positive word of mouth, and likelihood of repeating the experience in the future.

In the following section, the chapter illustrates the phenomenon of slow tourism. The third section describes outdoor tourism with a special focus on camping, while the fourth section introduces the concept of memorable tourism experience. The fifth section summarizes the research objectives. The sixth section presents the empirical study by illustrating the followed methodology and the results obtained from statistical analysis. The chapter concludes with a discussion of the empirical findings and their implications for scholars, managers, and policy-makers.

2. Slow Tourism

Slow tourism has emerged as a new form of tourism that is often considered a sustainable alternative to mass tourism (Fullagar, Markwell & Wilson, 2012). It has developed in response to recent environmental and societal issues, such as climate change, pollution, and stressful lifestyles, which make people feel powerless, out of control, and dehumanized (Frey & George, 2010; Howard, 2012; O'Regan, 2012; Parkins, 2004). More philosophically, the development of slow tourism could be seen as reflecting modern societies' increasing emphasis on an inner dimension of sustainability (Horlings, 2015; Jullien, 2018), which refers to all those personal beliefs and values (e.g., harmony with nature, universalism, etc.) that could hopefully lead individuals to change the world for the better by changing themselves. Slow tourism indeed shares the basic principles that drive similar phenomena, such as slow food and slow consumption (Petrini, 2001; Pietrykowski, 2004), which revolve around the idea that a decelerated lifestyle is more compatible with environmental sustainability, subjective happiness, and societal well-being (Hall, 2009; Lumsdon & McGrath, 2011).

Despite researchers' broad awareness of this topic, there is a lack of consensus in the literature about the conceptualization of slow tourism. While some scholars have defined slow tourism in terms of traveling, others have emphasized the motivational and attitudinal aspects of tourists' experiences. For instance, in the first strand of literature, Dickinson and Lumsdon (2010) defined it as a form of tourism in which tourists travel slowly from one destination to another, stay in a destination longer than usual, and consider the travel itself to be an essential part of their experience. As a result, slow tourism seems more compatible with environmental values (e.g., naturalness) and subjective wellness (e.g., relaxation) than other forms of tourism rooted in hasty traveling and a lack of time when visiting places (Moore, 2012). In the second stream of literature, Oh et al. (2016) proposed a conceptualization around the inner aspects of slow tourism, defining it as an experience in which tourists travel at a *mentally* slow pace—independently of the speed and manner of transportation—to satisfy intrinsic motivations and attain personal goals. According to Oh et al. (2016), the motivations for undertaking a slow tourism experience include *relaxation* (i.e., a tourist's desire to feel free of stress and anxiety), *self-reflection* (i.e., the tourist's desire to experience a sense of connection to the self), *escape* (i.e., a desire to feel disconnected from daily routines), *novelty-seeking* (i.e., a desire for newness and adventure when visiting new places), *engagement* (i.e., a desire to feel a sense of enjoyment and involvement when traveling), and *discovery* (i.e., a desire to derive pleasure from learning and understanding new things and ways of life). Personal goals connected to a slow tourism experience may include *revitalization* (i.e., the extent to which tourists feel reinvigorated and recharged after a holiday) and *self-enrichment* (i.e., the extent to which tourists are inspired and rediscover themselves) (Oh et al., 2016).

Slow tourism incorporates relevant experiential dimensions, such as sightseeing and immersing oneself in the local landscape (Heitmann, Robinson & Povey, 2011; Lumsdon & McGrath, 2011), interaction with residents and places (Dickinson & Lumsdon, 2010; Dickinson, Lumsdon & Robbins, 2011), spending time learning about places and local cultures (Moira et al., 2017), and living at a slower pace (Parkins & Craig, 2006). Recognizing this experiential component, many destinations such as Italy (Haupt, 2017) have been investing increasing amounts of resources to develop slow tourism and attract tourists with higher levels of spending power, environmental conscientiousness, and attention to authentic local cultures and traditions.

3. Outdoor Tourism and Camping

Outdoor tourism is a typical example of slow tourism (Dwyer & Edwards, 2000). It comprises those forms of tourism where most recreational activities occur in open-air environments. Outdoor tourism experiences typically incorporate adventure recreational activities that directly connect tourists to the natural environment, taking them away from stress, anxiety, and time pressure (Coghlan & Buckley, 2012; Newsome, Moore & Dowling, 2002; Weber, 2001). Such experiences include fishing, trekking, landscape watching, bicycling, and camping (Coghlan, 2007; Brooker & Joppe, 2013), which allow tourists to escape from daily life routines and savor a slower, more deliberate connection with nature.

Camping, the focal point of this study, is a typical outdoor tourism experience whereby tourists spend most of their time in wilderness or rural settings in strict connection with the natural environment (Brooker & Joppe, 2013). Thus, camping is a good example of slow tourism as it normally involves a decelerated life pace (Blichfeldt & Mikkelsen, 2015). This is not only because campers often lack access to modern high-speed technologies and services (e.g., fiber Internet connection, high-speed transportations) but also because they have opportunities to stay longer in campsites and devote enough time to discover natural attractions, play sport activities, and learn about local cultures and traditions, all while away from daily routines and stressful obligations.

Camping has witnessed substantial growth in recent times (Mikulić et al., 2017). In Italy, for instance, camping has developed substantially over the past few years in quantity and quality. In terms of quantity, Italy now claims 14% of all nights spent in campsites in the EU, which is the second-largest total in Europe (Eurostat, 2019). As for quality, Italian campsites have been increasingly recognized for the high quality of their infrastructures, services, and locations (*Il Sole 24 Ore*, 27 July 2017). In the wake of slow tourism, modern campsites have started offering differentiated and high-quality services that allow campers to live unique experiences, for example, by practicing sport activities and doing excursions along natural and historical attractions. Thanks to this evolution, camping is no longer considered an inexpensive temporary sojourn for tourists with limited budgets; rather, it is a niche tourism sector for certain tourists who seek authentic nature-based experiences, social interaction, and personal invigoration (Brooker & Joppe, 2014; Garst, Williams & Roggenbuck, 2010).

As camping has grown, so has scholarly interest in the topic. Past studies have investigated different strategic aspects of camping, such as campers' judgments of individual campsite attributes (Mikulić et al., 2017; Van Heerden, 2010), their satisfaction with campsites in general (Hardy, Ogunmokun & Winter, 2005; O'Neill, Riscinto & Hyfte, 2010), their perception of campsite pricing strategies (Park, Ellis, Kim & Prideaux, 2010), and how camping might serve regional repositioning purposes (Grzinic, Zarkovic & Zanketic, 2010). To date, only a few studies have investigated the experiential meanings that campers attribute to camping, but they generally support the idea that this form of tourism can contribute to human development and well-being. For instance, Garst et al. (2010) conducted a qualitative investigation on US campers that found that the most relevant meanings associated with this form of tourism are: *restoration* (i.e., relief from stress and anxiety), *special places* (i.e., places that evoke traditions and memories), *family functioning* (i.e., social and constructive interaction among family members), *experiencing nature* (i.e., staying in strict contact with the natural environment), *self-identity* (i.e., expressing self-related aspects), *social interaction* (i.e., socializing with others), and *children learning* (i.e., providing kids with learning opportunities). In a more general study on nature-based tourism, Breiby (2014) found that certain aesthetic components may enrich tourists' experiences. Specifically, her findings indicated that feeling a sense of harmony with nature and viewing beautiful landscapes are particularly relevant to fostering tourists' perceptions of extraordinary experiences.

Although interesting, the aforementioned findings do not provide a complete description of camping experiences. To satisfy this goal, we adopt and discuss the theoretical framework developed by Kim, Ritchie and McCormick (2012): the *memorable tourism experience* model.

4. Tourist Experience and Memorable Tourism Experience

Over the past few decades, modern economies have shifted from standardized production to customization (Breiby, 2014; Pine & Gilmore, 1999), whereby companies increasingly offer personalized solutions (i.e., products with integrated services). These solutions have often been designed around consumers' needs, which has had the effect of modifying their expectations and desires, as well as shaping their consumption experiences (Guido, 1992, 1998, 2014). Against this backdrop, scholars from different disciplines have committed more attention to studying *consumption experience* (Addis, 2005). In the marketing domain, some scholars have defined this construct as consumers' holistic perception about their whole interaction with a product, service, or brand, beyond the moment of mere purchase and usage (Guido, Bassi & Peluso, 2010; Peluso, 2011). Adopting a more psychological perspective, others have regarded consumption experience as a multidimensional concept that encompasses both rational and emotional aspects. For instance, in line with Holbrook and Hirschman's (1982) seminal work about the critical role of emotions in consumption (e.g., fun, pleasure), Schmitt (1999, 2011) looked at the consumption experience as a complex internal reaction that comprises sensory, emotional, cognitive, and behavioral responses. Building on this view, Brakus, Schmitt and Zarantonello (2009) developed a model that conceptualizes consumption experience with a product, service, or brand as a four-dimensional construct, which comprises a *sensory* dimension (i.e., the intensity with which a product, service, or brand stimulates senses), an *affective* dimension (i.e., the extent to which it arouses emotional feelings), a *behavioral* dimension (i.e., the extent to which it induces actions), and an *intellectual* dimension (i.e., the intensity with which it activates rational thinking).

An interesting contribution in this area of inquiry has been the distinction between *ordinary* and *extraordinary* experiences. While the former typically occur routinely within everyday life (e.g., eating), extraordinary experiences occur less frequently yet are far more emotionally intense, engaging, and transcendent, and thus completely immersive for the experiencing consumer (e.g., enjoying a holiday in a luxury resort or an excursion in a natural park) (Carù & Cova, 2006; Schmitt, 2011). Consumer research (Arnould & Price, 1993; Bhattacharjee & Mogilner, 2013; Lindberg & Østergaard, 2015) has found that extraordinary experiences can be humanistic (i.e., transformative for individuals) insofar as they can generate enduring memories, increase happiness, and stimulate personal growth.

The concept of extraordinary consumption experience is particularly suitable for describing tourism activities, which are experiential and exceptional in nature. Indeed, tourism research (e.g., Kim et al., 2012; Oh, Fiore & Jeong, 2007) has often drawn from the marketing literature to describe tourism experiences. In line with the notion of extraordinary experience, tourists' vacations often involve intensive activities that produce a feeling of complete immersion and, subsequently, long-last memories. Based on this reasoning, Kim et al. (2012) introduced the concept of *memorable tourism experience*, defined as an experience that tourists can positively remember for an indefinite time after it has occurred (see also Tung & Ritchie, 2011). Their work postulates that delivering *memorable tourism experiences* is critical to encouraging future revisits and positive word of mouth. Moreover, memories of such experiences seem particularly effective at fostering human development, as they can activate transformative processes that increase happiness and well-being (Sthapit & Coudounaris, 2018).

Of course, not all tourism experiences are unforgettable in a positive way; to activate good memories, they should have certain characteristics. Kim et al. (2012) empirically validated a model that conceptualizes *memorable tourism experience* through seven different dimensions that can independently contribute to unforgettable and extraordinary experiences. These seven dimensions included *hedonism* (i.e., the extent to which a tourist's experience is enjoyable and exciting), *novelty* (i.e., the extent to which the experience is perceived as unique), *local culture* (i.e., the extent to which it conveys positive impressions about local people and culture), *refreshment* (i.e., the extent to which it is perceived as liberating and revitalizing), *meaningfulness* (i.e., the extent to which it is self-relevant), *involvement* (i.e., the extent to which it meets the actual interests of tourists), and *knowledge* (i.e., the extent to which it allows tourists to enrich their knowledge).

5. Research Objectives

The foregoing discussion implies that there are positive consequences, both humanistically and managerially, to offering tourists the opportunity to live memorable experiences. From a humanistic perspective, such experiences are likely to increase personal happiness and well-being.

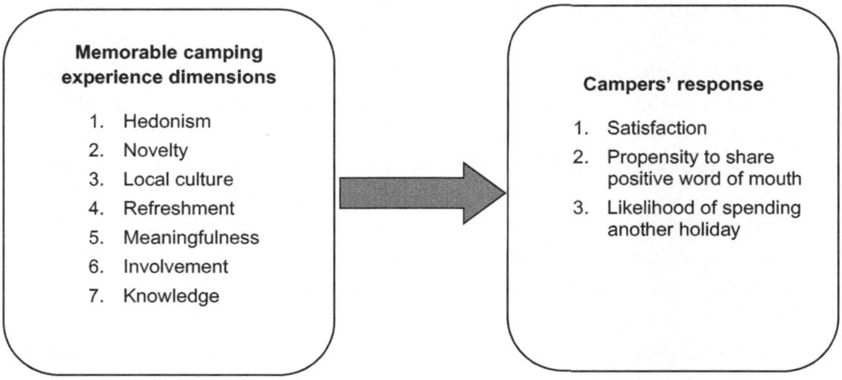

Figure 13.1 Theoretical framework
Source: Authors' own elaboration

Managerially, such experiences might lead to positive outcomes for the tourist attraction or destination, such as personal revisits or spreading positive information to others. Yet, despite increasing attention among scholars, the construct of *memorable tourism experience* has never been applied to a specific setting like camping. Therefore, the empirical study presented later in the chapter is the first to adopt Kim et al.'s (2012) model to describe camping experiences, with the final aim of contributing to the emerging humanist perspective on tourism.

Specifically, the study served two objectives: The first was to assess the dimensions of camping experience by adopting Kim et al.'s (2012) framework. In doing so, the study provides a better understanding of camping experiences and, more specifically, those dimensions that could render tourists' experiences at campsites as extraordinary and unforgettable. The second objective was to understand which experiential dimensions may lead to campers' positive responses in terms of satisfaction, positive word of mouth, and likelihood of repeating the experience in the future. Figure 13.1 summarizes the theoretical framework that underlies these objectives and the empirical study that follows.

6. Methodology

6.1 Sample Description

The study was conducted at a campsite in Apulia, South of Italy (Figure 13.2), which is among the top regions in Italy in terms of tourist campsite visits (ISTAT, 2016). The examined campsite is located on the

Figure 13.2 Apulia region, Italy

Source: Authors' own elaboration

west coast of Apulia and offers different services and accommodation solutions at differentiated price levels.

The data were collected by administering an online questionnaire to an initial sample of 589 campers, who were randomly selected from a list of campers who spent at least one vacation at the studied campsite. Among the 589 campers who received the questionnaire via e-mail, 438 (65% males, 35% females; M_{Age} = 49.23, SD_{Age} = 11.25) participated in the survey (response rate = 74.36%). These 438 campers constituted the final sample, although 15 were subsequently removed from some statistical analyses because of skipping at least one question regarding the constructs of interest. The sample included campers with a different number of prior experiences with the campsite: 38% of respondents reported that they had only stayed one time in that campsite before, 14% reported that they had stayed two times, 14% three or four times, 18% five to ten times, and 16% more than ten times. Most respondents (67%) reported that they had stayed there with family (or their partner), 29% stayed with both family and friends, 3% with friends, and 1% stayed alone. Regarding the area of residence, 86% of respondents were Italian, and 14% were international tourists. Among the Italian respondents, 63% lived in Northern Italy, 13% lived in Central Italy, and 24% in Southern Italy. Among the international respondents, 74% came from EU countries and 26% from non-EU countries.

6.2 The Questionnaire

The questionnaire featured four sections. The first section comprised a set of introductory questions asking respondents to indicate: how often they had previously stayed at that campsite for holiday reasons, using a five-point scale (1 = one time, 2 = two times, 3 = three or four times, 4 = five to ten times, 5 = more than ten times); and with whom they usually stayed there, using a four-point nominal scale (1 = I stayed alone, 2 = with my family, 3 = with friends, 4 = with both family and friends).

The second section of the questionnaire included an adapted version of the *memorable tourism experience* scale (Kim et al., 2012), which comprised 25 statements regarding the respondents' experience with the campsite. Respondents were asked to indicate their degree of agreement with each of those statements on a seven-point Likert scale (1 = completely disagree, 7 = completely agree). Table 13.1 reports the list of those statements, associated with the seven dimensions of the assessed construct, along with descriptive statistics computed for each item. For each dimension, Table 13.1 also reports the Cronbach's α index, which is a measure of the degree of consistency among the respective items. Cronbach's α indices were all higher than 0.8; thus, the items assessing each of the seven dimensions could be averaged to obtain a composite measure of that dimension.

The third section of the questionnaire included one question regarding respondents' overall satisfaction with their camping experience, which was adapted from Mägi (2003) and assessed on an 11-point scale (i.e., "How satisfied are you with your whole experience during your stay at that campsite?"; 0 = not at all satisfied, 10 = completely satisfied). Three questions assessed the respondents' tendency to share positive information about their camping experiences. The first question was derived from Reichheld (2003) and assessed recommendation likelihood on an 11-point scale (i.e., "How likely is it that you will recommend a holiday at that campsite to your friends, relatives, and/or colleagues?"; 0 = very unlikely, 10 = very likely). Two additional questions, assessed on an 11-point scale, supplemented the former by measuring respondents'

Table 13.1 Adapted version of the memorable tourism experience scale

Dimension (Cronbach's α)	*Item*	*M*	*SD*
Hedonism (0.85)	1. I felt thrilled about having a new experience	5.01	1.56
	2. I indulged in the activities	4.90	1.59
	3. I really enjoyed this tourism experience	5.40	1.42
	4. The holiday at that campsite was exciting	5.18	1.52
Novelty (0.87)	5. It was a once-in-a-lifetime experience	4.81	1.65
	6. It was unique	4.65	1.76
	7. It was different from previous experiences	4.87	1.68
	8. I experienced something new (e.g., food, activities, etc.)	4.52	1.79
Local culture (0.89)	9. I had good impressions about local people	5.66	1.45
	10. I appreciated the authentic products and traditions of the place	5.70	1.46
	11. I closely experienced the local culture	5.04	1.58
	12. Local people were friendly	5.80	1.37
Refreshment (0.95)	13. The holiday at that campsite was liberating	5.90	1.45
	14. I enjoyed a sense of freedom	5.99	1.40
	15. The holiday was refreshing	6.01	1.39
	16. It was revitalizing	5.80	1.47
Meaningfulness (0.83)	17. I did something meaningful	5.21	1.62
	18. I did something important	5.13	1.66
	19. I learned about myself	3.58	1.90
Involvement (0.84)	20. I visited a place where I really wanted to go	5.18	1.71
	21. I enjoyed activities which I really wanted to do	4.96	1.75
	22. I was interested in the main activities of that tourism experience	4.51	1.75
Knowledge (0.88)	23. I explored new things	4.43	1.68
	24. I learned more about new things and life	4.08	1.72
	25. I experienced a new culture	4.12	1.90

N = 438.

Source: Authors' own elaboration

propensity to share electronic word of mouth (i.e., "How often have you shared digital contents, such as photos or video, regarding your stay at that campsite on social media like Facebook, Instagram, Twitter, or Whatsapp?"; "How often have you written reviews regarding your stay at that campsite on the Internet?"; 0 = never, 10 = very often). As these three items reached an acceptable level of internal consistency ($\alpha = 0.70$), their scores were aggregated to obtain a composite measure of respondents' propensity to share positive word of mouth. This section of the questionnaire also included a question adapted from Morgan and Rego (2006), which asked respondents to indicate, on an 11-point scale, their degree of loyalty toward the campsite, in terms of their likelihood of spending another holiday at the campsite (i.e., "How likely is it that you will spend another holiday at that campsite during the next two years?"; 0 = very unlikely, 10 = very likely). The fourth section concluded the questionnaire with socio-demographic questions regarding respondents' age, gender, and area of residence.

6.3 Statistical Analysis

We began the statistical analysis by generating descriptive statistics for the assessed variables. First, we averaged the items assessing each of the seven dimensions characterizing the *memorable tourism experience* construct to obtain a composite measure of that dimension. The descriptive statistics computed at a dimensional level indicated that *refreshment* ($M = 5.93$, $SD = 1.34$) and *local culture* ($M = 5.48$, $SD = 1.30$) were the most relevant dimensions. *Hedonism* ($M = 5.12$, $SD = 1.27$), *involvement* ($M = 4.88$, $SD = 1.51$), *novelty* ($M = 4.72$, $SD = 1.46$), and *meaningfulness* ($M = 4.64$, $SD = 1.49$) were moderately relevant. *Knowledge* was the least relevant ($M = 4.21$, $SD = 1.59$). These preliminary findings suggest that respondents perceived their camping experience as an opportunity to feel a sense of freedom or revitalization, as well as to immerse themselves in local cultures and traditions.

We also computed descriptive statistics regarding respondents' degree of satisfaction, propensity to share positive word of mouth, and likelihood of spending another holiday at the campsite. The results indicated that, on average, respondents were very satisfied with their camping experiences ($M = 9.31$; $SD = 1.85$), moderately inclined to share positive word of mouth ($M = 6.76$; $SD = 2.19$), and likely to spend another holiday at the campsite over the subsequent two years ($M = 8.95$; $SD = 2.61$).

As a second step of the statistical analysis, we computed bivariate correlations among all variables. The results summarized in Table 13.2 show that all correlation coefficients were positive and significant ($ps < 0.001$). Thus, we adopted a regression-based analytical approach to understand which of the seven dimensions characterizing *memorable tourism experience* were critical to determining respondents' reactions.

Specifically, we estimated the linear relationship among each of the seven dimensions of *memorable tourism experience*, on one hand, and each of three measures (respondents' satisfaction, positive word of mouth, and likelihood to revisit), on the other hand. In total, we conducted three multiple regression analyses using the seven *memorable tourism experience* dimensions as independent variables. Each of the three analyses was accompanied by one dependent variable—respectively, respondents' satisfaction, propensity to share positive word of mouth, and likelihood of revisiting. Table 13.3 summarizes the results of the three regression analyses.

In the first analysis, we regressed respondents' degree of satisfaction on the seven *memorable tourism experience* dimensions. The results showed that *hedonism* ($b = 0.40$, $p < 0.001$), *local culture* ($b = 0.14$,

Table 13.2 Bivariate correlations among relevant variables

Var.	*HED*	*NOV*	*LC*	*REF*	*MEA*	*INV*	*KNO*	*SOD*	*PWOM*	*LIK*
HED	1.00									
NOV	0.79	1.00								
LC	0.60	0.62	1.00							
REF	0.69	0.66	0.64	1.00						
MEA	0.64	0.71	0.58	0.64	1.00					
INV	0.69	0.66	0.64	0.62	0.66	1.00				
KNO	0.56	0.66	0.59	0.46	0.66	0.70	1.00			
SOD	0.71[a]	0.64[a]	0.59[a]	0.77[a]	0.55[a]	0.59[a]	0.46[a]	1.00		
PWOM	0.48[a]	0.48[a]	0.43[a]	0.41[a]	0.43[a]	0.41[a]	0.41[a]	0.46[a]	1.00	
LIK	0.50[a]	0.46[a]	0.46[a]	0.57[a]	0.45[a]	0.46[a]	0.33[a]	0.63[a]	0.50[a]	1.00

N = 438. [a] *N* = 423 (15 cases were removed from the sample of 438 respondents because of missing values on the corresponding variables). HED = Hedonism; NOV = Novely; LC = Local culture; REF = Refreshment; MEA = Meaningfulness; INV = Involvement; KNO = Knowledge; SOD = Satisfaction; PWOM = Propensity to share positive word of mouth (WOM); LIK = Likelihood of spending another holiday at the campsite. All correlations are significant at a 0.001 level.

Source: Authors' own elaboration

Table 13.3 Results of the regression analyses

Variable	*Dependent variable: Degree of satisfaction*		*Dependent variable: Propensity to share positive WOM*		*Dependent variable: Likelihood of spending another holiday*	
	b	*Std. error*	*b*	*Std. error*	*b*	*Std. error*
Constant	2.00***	0.28	1.56**	0.47	1.34*	0.53
Hedonism	0.40***	0.08	0.36**	0.13	0.20	0.15
Novelty	0.06	0.07	0.15	0.12	0.04	0.13
Local culture	0.14*	0.06	0.25*	0.10	0.23*	0.12
Refreshment	0.71***	0.06	0.06	0.11	0.69***	0.13
Meaningfulness	−0.08	0.06	0.12	0.10	0.13	0.11
Involvement	0.09	0.06	−0.04	0.10	0.22	0.11
Knowledge	−0.02	0.05	0.12	0.09	−0.15	0.10
R^2 (Adj. R^2)	0.66 (0.66)		0.29 (0.28)		0.36 (0.35)	
N[a]	423		423		423	

[a] A total of 15 cases were removed from the sample of 438 respondents because of missing values on the corresponding variables. * significance at a 0.05 level; ** significance at a 0.01 level; *** significance at a 0.001 level.

Source: Authors' own elaboration

$p < 0.05$), and *refreshment* ($b = 0.71$, $p < 0.001$) were positively related to satisfaction. In the second analysis, we regressed respondents' propensity to share positive word of mouth on those dimensions, showing that *hedonism* ($b = 0.36$, $p < 0.01$) and *local culture* ($b = 0.25$, $p < 0.05$)

were positively related to positive word of mouth. In the third analysis, we regressed respondents' likelihood of spending another holiday at the campsite on the seven *memorable tourism experience* dimensions, showing that *local culture* ($b = 0.23$, $p = 0.05$) and *refreshment* ($b = 0.69$, $p < 0.001$) were positively related to that dependent variable.

7. General Discussion and Conclusion

This chapter presented an empirical study that investigated the factors that transform campers' holidays into memorable tourism experiences. To this end, we adopted Kim et al.'s (2012) model to assess the critical dimensions and then estimated the relationship between each dimension and campers' reaction in terms of satisfaction, positive word of mouth, and likelihood of repeating the experience. According to the results, the main aspects that render a camping experience as memorable are a sense of refreshment in campers, a strict interaction with local people and culture, and a general feeling of enjoyment. These three dimensions are also critical to increasing campers' satisfaction, propensity for sharing positive word of mouth, and likelihood of revisiting the campsite during a future holiday.

7.1 Implications

From a theoretical perspective, this work is the first to look at *camping* from a humanistic perspective—that is, as a common form of slow tourism that has the potential to promote human development. Camping does indeed allow tourists to experience time and space at a decelerated pace and in strict connection with the natural environment and local people, away from the fast and stressful pace of daily routine. Camping experiences can sometimes be unforgettable and extraordinary (Arnould & Price, 1993; Bhattacharjee & Mogilner, 2013), which can then activate internal transformative processes that lead individual campers to change themselves for the better. The results support the idea that camping experiences are memorable, especially when they convey a sense of refreshment, allow campers to interact with local people and culture, and promote enjoyment.

From an operational perspective, this study has implications for managers and policy-makers. Campsite managers could leverage our findings to deliver unique and unforgettable camping experiences. For instance, campsite managers should convey in campers a sense of refreshment and freedom; give them the opportunity to interact with local people, traditions, and culture; and try to ensure their enjoyment and excitement. Operationally, managers could organize their campsites and differentiate their services in accordance with the aforementioned goals. Meanwhile, policy-makers might leverage our results in redesigning and repositioning tourist destinations. Camping tourism has clear sustainability and humanistic connotations, insofar as campers can live in destinations at a slow pace and in close connection with the natural environment, the local culture, and the self. Therefore, policy-makers interested in fostering a sustainable and humanistic development of tourist destinations could invest resources into camping promotions. Of course, this does not mean that policy-makers should allow campsites to proliferate without control; rather, they should do their best to create good conditions—in terms of infrastructures and regulations—that preserve the inherent connotations and humanistic nature of camping.

7.2 Limitations and Future Research Directions

The study presented in this chapter has some limitations that offer opportunities for future research. First, the data were collected from a single campsite in Apulia (Southern Italy). Therefore, notwithstanding their statistical significance, the results should be generalized with caution. Future studies could replicate the study in other campsites in other geographical areas; they could also verify whether the findings obtained in this study are generalizable to other forms of slow tourism (e.g., summer homes or cottages).

Second, the study did not analyze the role of individual differences among campers, which might moderate the relationship between the experiential dimensions of camping and campers' reactions. Future studies could do this by assessing personality traits, values, cultural background, and the like, and then using such measures as moderators in the statistical analysis. They could also collect data from multiple campsites to examine the potential moderating effects of campsite-related differences such as geographical location, type of services offered, and so on.

Finally, future studies could combine quantitative methods with qualitative research techniques (e.g., in-depth interviews) to gain a deeper understanding of campers' experiences. Such research might produce new insights about how people actually connect camping to the values of freedom, happiness, nature, sociality, and culture.

References

Addis, M. (2005). *L'esperienza di consumo: Analisi e prospettive di marketing*. Milan: Pearson.

Arnould, E. J., & Price, L. L. (1993). River magic: Extraordinary experience and the extended service encounter. *Journal of Consumer Research*, *20*(1), 24–45.

Bhattacharjee, A., & Mogilner, C. (2013). Happiness from ordinary and extraordinary experiences. *Journal of Consumer Research*, *41*(1), 1–17.

Blichfeldt, B. S., & Mikkelsen, M. V. (2015). Camping tourism. In J. Jafari & H. Xiao (Eds.), *Encyclopedia of tourism* (pp. 1–2). New York: Springer.

Botterill, D. T. (1989). Humanistic tourism? Personal construction of a tourist: Sam visits Japan. *Leisure Studies*, *8*(3), 281–293. doi:10.1080/02614368900390281

Brakus, J. J., Schmitt, J. H., & Zarantonello, L. (2009). Brand experience: What is it? How is it measured? Does it affect loyalty? *Journal of Marketing*, *73*(3), 52–68.

Breiby, M. A. (2014). Exploring aesthetic dimensions in a nature-based tourism context. *Journal of Vacation Marketing*, *20*(2), 163–173.

Brooker, E., & Joppe, M. (2013). Trends in camping and outdoor hospitality—An international review. *Journal of Outdoor Recreation and Tourism*, *3*, 1–6. doi:10.1016/j.jort.2013.04.005

Brooker, E., & Joppe, M. (2014). A critical review of camping research and direction for future studies. *Journal of Vacation Marketing*, *20*(4), 335–351.

Carù, A., & Cova, B. (2006). How to facilitate immersion in a consumption experience: Appropriation operations and service elements. *Journal of Consumer Behaviour: An International Research Review*, *5*(1), 4–14.

Caton, K. (2016). A humanist paradigm for tourism studies? Envisioning a collective alternative to epistemic literalism. In A. M. Munar & T. Jamal (Eds.), *Tourism research paradigms: Critical and emergent knowledges* (Vol. 22, pp. 35–56). Bingley: Emerald.

Coghlan, A. (2007). Towards an integrated image-based typology of volunteer tourism organizations. *Journal of Sustainable Tourism*, *15*(3), 267–287.

Coghlan, A., & Buckley, R. C. (2012). Nature-based tourism. In A. Holden & D. Fennell (Eds.), *The Routledge handbook of tourism and the environment* (pp. 334–344). London: Routledge.

Dickinson, J. E., & Lumsdon, L. M. (2010). *Slow travel and tourism*. London: Earthscan.

Dickinson, J. E., Lumsdon, L. M., & Robbins, D. (2011). Slow travel: Issues for tourism and climate change. *Journal of Sustainable Tourism*, *19*(3), 281–300.

Dwyer, L., & Edwards, D. (2000). Nature-based tourism on the edge of urban development. *Journal of Sustainable Tourism*, *8*(4), 267–287.

Eurostat. (2019). *Thinking of going camping this summer?* Retrieved July 16, 2019, from https://ec.europa.eu/eurostat/web/products-eurostat-news/-/DDN-20190613-1

Frey, N., & George, R. (2010). Responsible tourism management: The missing link between business owners' attitudes and behaviour in the cape town tourism industry. *Tourism Management*, *31*(5), 621–628.

Fullagar, S., Markwell, K., & Wilson, E. (2012). *Slow tourism: Experiences and mobilities*. Bristol: Channel View Publications.

Garst, B. A., Williams, D. R., & Roggenbuck, J. W. (2010). Exploring early twenty-first century developed forest camping experiences and meanings. *Leisure Sciences*, *32*(1), 90–107.

Grzinic, J., Zarkovic, A., & Zanketic, P. (2010). Positioning of tourism in central Dalmatia through the development of camping tourism. *International Journal of Economic Perspectives*, *4*(3), 525–535.

Guido, G. (1992). What U.S. marketers should consider in planning a pan-European approach. *Journal of Consumer Marketing*, *9*(2), 29–33.

Guido, G. (1998). Consumers as products: An alternative approach to the customer-orientation concept. In D. Kantarelis (Ed.), *Business and economics for the 21th century* (Vol. 2, pp. 39–45). Worcester: Business & Economics Society International.

Guido, G. (2014). Customer satisfaction. In C. L. Cooper (Ed.), *Wiley encyclopedia of management, Marketing* (Vol. 9, pp. 139–147). New York: Wiley.

Guido, G., Bassi, F., & Peluso, A. M. (2010). *La soddisfazione del consumatore: la misura della customer satisfaction nelle esperienze di consumo*. Milan: Franco Angeli.

Hall, C. M. (2009). Degrowing tourism: Décroissance, sustainable consumption and steady-state tourism. *Anatolia: An International Journal of Tourism and Hospitality Research*, *20*(1), 46–61.

Hardy, T., Ogunmokun, G., & Winter, C. (2005). An exploratory study of factors influencing campers, level of loyalty to camping sites in the tourism industry. In *Proceedings of the 19th Australian and New Zealand academy of management conference* (pp. 1–11). Canberra, Australia: Australian and New Zealand Academy of Management.

Haupt, T. (2017). Italy to focus more on slow tourism. *Tourism Review News*. Retrieved July 16, 2019, from www.tourism-review.com/slow-tourism-is-the-trend-in-italy-news10372

Heitmann, S., Robinson, P., & Povey, G. (2011). Slow food, slow cities, and slow tourism. In P. Robinson, S. Heitmann, & P. Dieke (Eds.), *Research themes for tourism* (pp. 114–127). London: MPG Books Group.

Holbrook, M. B., & Hirschman, E. C. (1982). The experiential aspects of consumption: Consumer fantasies, feelings, and fun. *Journal of Consumer Research*, *9*(2), 132–140.

Horlings, L. G. (2015). The inner dimension of sustainability: Personal and cultural values. *Current Opinion in Environmental Sustainability*, *14*, 163–169. doi:10.1016/j.cosust.2015.06.006

Howard, C. (2012). Speeding up and slowing down: Pilgrimage and slow travel through time. In S. Fullagar, K. Markwell, & E. Wilson (Eds.), *Slow tourism: Experiences and mobilities* (pp. 11–24). Bristol: Channel View Publications.

Il Sole 24 Ore. (2017). *Campeggi al top in Italia. Le strutture premiate per sport, ristorazione, wellness*. Retrieved July 16, 2019, from www.ilsole24ore.com/art/campeggi-top-italia-strutture-premiate-sport-ristorazione-wellness-AE5W6c3B?refresh_ce=1

ISTAT. (2016). *Presenze nei campeggi e villaggi turistici in Italia per regione*. Retrieved July 16, 2019, from www.datiopen.it/it/opendata/Presenze_nei_campeggi_e_villaggi_turistici_in_italia_per_regione

Jullien, F. (2018). *Living off landscape, or the un-thought of in reason*. Lanham, MD: Rowman & Littlefield International.

Kim, J. H., Ritchie, J. B., & McCormick, B. (2012). Development of a scale to measure memorable tourism experiences. *Journal of Travel Research*, *51*(1), 12–25.

Lindberg, F., & Østergaard, P. (2015). Extraordinary consumer experiences: Why immersion and transformation cause trouble. *Journal of Consumer Behaviour*, *14*(4), 248–260.

Lumsdon, L. M., & McGrath, P. (2011). Developing a conceptual framework for slow travel: A grounded theory approach. *Journal of Sustainable Tourism*, *19*(3), 265–279.

Mägi, A. W. (2003). Share of wallet in retailing: The effects of customer satisfaction, loyalty cards and shopper characteristics. *Journal of Retailing*, *79*, 97–106. doi:10.1016/S0022-4359(03)00008-3

Melé, D. (2003). The challenge of humanistic management. *Journal of Business Ethics*, *44*, 77–88. doi:10.1023/A:1023298710412

Mikulić, J., Prebežac, D., Šerić, M., & Krešić, D. (2017). Campsite choice and the camping tourism experience: Investigating decisive campsite attributes using relevance-determinance analysis. *Tourism Management*, *59*, 226–233. doi:10.1016/j.tourman.2016.07.020

Moira, P., Mylonopoulos, D., & Kondoudaki, A. (2017). The application of slow movement to tourism: Is slow tourism a new paradigm? *Journal of Tourism and Leisure Studies*, *2*(2), 1–10.

Moore, K. (2012). On the periphery of pleasure: Hedonics, eudaimonics, and slow travel. In S. Fullagar, K. Markwell, & E. Wilson (Eds.), *Slow tourism: Experiences and mobilities* (pp. 25–35). Bristol: Channel View Publications.

Morgan, N. A., & Rego, L. L. (2006). The value of different customer satisfaction and loyalty metrics in predicting business performance. *Marketing Science*, *25*(5), 426–439.

Newsome, D., Moore, S. A., & Dowling, R. K. (2002). *Natural area tourism: Ecology, impacts and management*. Clevedon and Bristol: Channel View Publications.

Oh, H., Assaf, A. G., & Baloglu, S. (2016). Motivations and goals of slow tourism. *Journal of Travel Research*, *55*(2), 205–219.

Oh, H., Fiore, A. M., & Jeong, M. (2007). Measuring experience economy concepts: Tourism applications. *Journal of Travel Research*, *46*(2), 119–132.

O'Neill, M., Riscinto, K., & Hyfte, M. (2010). Defining visitor satisfaction in the context of camping oriented nature based tourism—The driving force of quality. *Journal of Vacation Marketing*, *16*(2), 141–156.

O'Regan, M. (2012). Alternative mobility cultures and the resurgence of hitch—hiking. In S. Fullagar, K. Markwell, & E. Wilson (Eds.), *Slow tourism: Experiences and mobilities* (pp. 128–142). Bristol: Channel View Publications.

Park, J., Ellis, G. D., Kim, S. S., & Prideaux, B. (2010). An investigation of perceptions of social equity and price acceptability judgments for campers in the U.S. national forest. *Tourism Management*, *31*(2), 202–212.

Parkins, W. (2004). Out of time: Fast subjects and slow living. *Time and Society*, *13*(2–3), 363–382.

Parkins, W., & Craig, G. (2006). *Slow living*. Oxford: Berg Publishers.

Peluso, A. M. (2011). *Consumer satisfaction: Advancements in theory, modeling, and empirical findings*. Bern: Peter Lang.

Petrini, C. (2001). *Slow food: The case for taste*. New York: Columbia University Press.

Pietrykowski, B. (2004). You are what you eat: The social economy of the slow food movement. *Review of Social Economy*, *62*(3), 307–321.

Pine, B. J., & Gilmore, J. H. (1999). *The experience economy: Work is theatre & every business a stage*. Boston: Harvard Business Press.

Reichheld, F. F. (2003). The one number you need to grow. *Harvard Business Review*, *81*(12), 46–55.

Schmitt, B. (1999). *Experiential marketing*. New York: Free Press.

Schmitt, B. (2011). Experience marketing: Concepts, frameworks and consumer insights. *Foundations and Trends in Marketing*, *5*(2), 55–112.

Sthapit, E., & Coudounaris, D. N. (2018). Memorable tourism experiences: Antecedents and outcomes. *Scandinavian Journal of Hospitality and Tourism*, *18*(1), 72–94.

Tung, V. W. S., & Ritchie, J. R. B. (2011). Exploring the essence of memorable tourism experiences. *Annals of Tourism Research*, *38*(4), 1367–1386.

Van Heerden, C. H. (2010). An exploratory analysis of leisure caravanning in the Kruger national park in South Africa. *Innovative Marketing*, 6(1), 66–72.

Weber, K. (2001). Outdoor adventure tourism: A review of research approaches. *Annals of Tourism Research*, *28*(2), 360–377.

Part IV
Conclusions

14 Lessons for Shared Value Creation in Tourism

The Pandemic Challenge

Maria Della Lucia, Ernestina Giudici, and Frédéric Dimanche

1. Introduction

A crisis is an unexpected and disruptive event entailing a progressive process that may entangle rapidly, triggers other crises, and has extended consequences (Hart, Heyse & Boin, 2001). The crisis may be activated by a combination of factors—from environmental forces or technological failure (Perrow, 1999) to human error (Reason, 1990)—and entails a transformation of the old system, which needs a qualitative change (Venette, 2003). If change is not needed, the event could more accurately be described as a system failure or incident. Crisis management is a process entailing several levels of activity (prevention, planning, training, response, and recovery) by which an organization deals with disruptive events that threaten to harm the organization, its stakeholders, or the general public (Mikušová & Horváthová, 2019). Identifying a sector or an organization's vulnerabilities is essential for crisis management. A crisis is not only a failure or a threat; it is also an opportunity for change and for improvement (Ioannides & Gyimóthy, 2020). This is particularly true for tourism today.

The COVID-19 pandemic has been defined as the most significant crisis of our generation (Harari, 2020). Its impact on our life as humans, citizens, workers, consumers, and travelers is probably unprecedented since World War II (Gössling, Scott & Hall, 2020). Travel and tourism as a single global, pervasive, and cross-sectoral industry has been deeply affected because of interdependences among systems (Della Lucia, 2013)—primarily healthcare, the economy, politics and governance, culture, and the environment. Most symptoms resulting from the disease shown by the tourism systems—and its related treatments and responses—have highlighted unsustainable aspects of our globalized and consumer-driven world. On the positive side of the ledger, there has been a short-term relief and recovery of tourism-related natural ecosystems and wildlife (Gardiner, 2020).

In the surge of academic contributions published since the start of the crisis, there is a stream bringing forward and addressing—in the light of COVID-19—topics that still need critical discussion. These revolve around sustainability and its dimensions—the economy, society, the environment, and governance—as well as humanistic management (e.g. ATLAS, 2020; Della Lucia & Giudici, 2021).

This chapter discusses the COVID-19 pandemic–induced impact on tourism in a critical perspective and traces future scenarios: tourism as usual, the new normal, and *humanistic tourism*. Recommendations for building humanistic tourism are offered. They are based on shared value processes (Porter & Kramer, 2012) at the interplay between sustainability and humanistic management, and they revolve around four clusters of interactions: *human vs. human, human vs. nature, human vs. technology, and human vs. the economy*. In the conclusion, the poetic fable "The Tin Forest" encourages us to act in manners favoring the emergence of a different world.

2. Impact of the COVID-19 Pandemic on Tourism: Uncovering Uneven Systems

A worldwide pandemic was declared on March 11, 2020, by the World Health Organization. As a consequence, the global tourism industry came to a halt as a potential vector of the virus, becoming a victim of the spread of infectious diseases (Novelli, 2020). With no vaccine to prevent the disease and limited medical treatments, most governments responded with short-term emergency measures. These clustered around the need to prioritize the population's health and safety over the economy either by adopting strict confinement regulations (e.g. Italy, France) or through looser approaches (e.g. UK, United States). Lockdown (home isolation, quarantine), social distancing, the closure of schools and universities, as well as non-essential businesses and workplaces, bans on gatherings of people over certain numbers, cancelling or postponing any kind of major events (political, sports, leisure, etc.), and the closure of borders (Gössling et al., 2020) all affected tourism systems, which are important components of national economies. At the same time, these measures curtailed personal freedoms, real and perceived, in work, civic life, and leisure (Havitz, Pritchard & Dimanche, 2020). Because of this reduced freedom, people in some countries argued that individuals, not governments, should decide whether "to exercise their right to work, to worship and to play. Or to even stay at home" (Freeman, 2020, April 15). The outbreak and the related contagion containment measures have positively affected the environment and the climate but have also provided cover for illegal activities, hindered environmental diplomacy efforts, and supported pro-growth policies, strategies, and investments (Geneva Environment Network, 2020). The pandemic has highlighted the magnitude and scope of tourism's global importance as well as its vulnerability.

In 2019, international tourism arrivals reached 1.5 billion, and tourism represented 10.3% of the global economy ($8.9 trillion), employing 330 million people (WTTC, 2020). The halt, for several months, and the consequent slowdown of international and domestic tourism, which has affected air and land transport, cruises, accommodations, cafés and restaurants, conventions, festivals, meetings, and sports events, has had a deep impact on 2020 tourism performances. According to WTTC's (2020) best- and worst-case scenarios, global domestic and international arrivals were expected to decrease by 26% (62% in the worst case) and 41% (73%), respectively, compared with 2019. An assumption was made that current restrictions would start to ease by June (September in the worst case) for domestic travel and short-haul/regional and by August (November in the worst case) for intercontinental travel. The global loss of travel and tourism ranges from $2.7 to $5.5 billion in GDP and from 98.2 to 197.5 million in terms of jobs, based on the best/worst case scenario, respectively. Total economic impact is an estimate of the effect on sales, production, income, added value, and employment of the demand for goods and services from visitors and tourism businesses through the sectorial interdependences of the economic system (Della Lucia, 2013). Spending on goods and services (direct effects) by visitors (e.g. overnight stays, meals, shopping, tickets) and tourism businesses (e.g. furniture, equipment, advertisement, rentals, fees) has multiplicative indirect and induced effects. The indirect multiplicative effects are related to the production involved in supplying the input needed to provide the goods and services required (e.g. hotel, restaurant, shops, attractions, transport). The induced multiplicative effect is the increased consumption of residents and the improved standard of living made possible by incomes generated, directly or indirectly, through tourism.

Because of the complexity of the tourism system (Della Lucia, Giudici & Secchi, 2021), the pandemic triggered de-multiplicative (economic) processes. The pandemic affected virtually all components of the tourism and hospitality value chain embedded in different models of tourism development (Weaver, 2000), through sectoral interdependences but also social and institutional ones. The impact of the pandemic is felt in mainstream tourism destinations strongly tied to tourism and international travel (Dodds & Butler, 2019; Seraphin, Sheeran & Pilato, 2018)—in fact, sustainable and unsustainable tourism destinations, which were suffering from overtourism, are now experiencing undertourism. Even community-based destinations are suffering from a great decline in small-scale tourism, which puts in question the fragile economic and environmental balance fostered by effective quality regulation (Weaver, 2012). For example, in destinations focusing on wildlife and nature tourism (Hockings et al., 2020), the loss of income from tourism endangers indigenous peoples and local communities in and around protected and conservation areas, which are dependent on tourism-derived income. This is threatening their livelihoods and, in turn, undermining the functioning of ecosystem processes and services, as well as their very identity and culture.

From the demand side, uncertainty, health risk, financial reasons, and government-imposed restrictions affected, and will continue to affect in the near future, the trust and confidence needed for people to be engaged in tourism of any kind (Havitz et al., 2020). Our society "on the move" has become accustomed to taking mobility and tourism far away from home for granted. Being capable of effectively handling tensions between the "freedom to act" and the freedom "from constraints," along with proper balancing between perceived and actual freedom, has become vital for human flourishing and the shaping of social identity (Havitz et al., 2020). The freedom (Ellis & Witt, 1984) to develop personal meaning and self-responsibility are at the core of the intention to move and encounter "others" elsewhere (Pritchard, Havitz & Howard, 1999).

Because of interdependences of a different kind, global tourism systems are currently suffering and are metaphorically on life support. Small, private stakeholders tend to be more fragile, and some will not survive the COVID-19–induced crisis. Responses from governments to COVID-19 have shown an unprecedented level and speed of policy and legislative action (Higgins-Desbiolles, 2020b), but they merely contribute to maintaining a system that has shown its limits. They are indeed putting in place economic support strategies to help large travel and tourism companies, small entrepreneurs, and communities survive the crisis. However, in so doing, many countries are taking on significant deficit financing and/or reviewing their spending priorities as a consequence of the radical level of indebtedness and deficit spending. These policies raise arguments about the rationales or norms (Weaver, 2012)—pro-growth, sustainability-conducive regulation, or hybrid norm—that are followed to allocate resources to tourism systems, to re-allocate to this pandemic economic response the budgets originally dedicated to other policies (e.g. environmental protection), and to boost the economy for financial gains in the short- to mid-range terms.

In emergency situations, the rollbacks concerning environmental protections that govern clean air, water, and toxic chemicals—among others—is not only a risk. For example, more than 60 environmental rules and regulations were officially reversed, revoked, or otherwise rolled back under the Trump administration, and an additional 34 rollbacks are still being considered (Popovich, Albeck-Ripka & Pierre-Louis, 2020). Legal efforts to downgrade the protection given to conservation areas, to reduce their size, or even to degazette them entirely have been proposed or enacted in many countries (Mascia & Pailler, 2011). The legal encouragement of deforestation, fragmentation, and ecosystem disruption has led to a greater risk of infectious zoonotic diseases. This goes along with illegal activities such as deforestation of the Amazon rainforest and poaching in Africa (Hockings et al., 2020), that make natural areas and wildlife even more vulnerable. In addition, the dramatic improvements in air and water quality, that were observed because of the lockdown, are, of course, temporary. Pollution and other environmental problems return to pre-COVID levels as soon as the restrictions on travel, production, and consumption are released (Gardiner, 2020). The lobbying efforts made to secure advantages—cash, regulatory rollbacks, and other special favors—made by powerful industries (e.g. fossil fuels, plastics, airlines, and automobiles) and some governments, to foster recovery and overcome the recession, could affect the environment even more. The current likely hypothesis is that dominant forces will try to exploit the pandemic crisis as an excuse to implement the by-now familiar disastrous capitalist playbook, pushed through further privatizations and corporate consolidation (Fletcher, Murray Mas, Blázquez-Salom & Blanco-Romero, 2020).

3. Future Tourism Scenarios

Under conditions of uncertainty, it is early for a comprehensive analysis of the pandemic's repercussions on the political-economic structure of tourism going forward (Bianchi, 2020). As a result, recent contributions dealing with the impact of COVID-19 on tourism, the paradoxes of the growth model of tourism, and the need to rebuild tourism, have drawn potential scenarios for the future of tourism (ATLAS, 2020; Brouder, 2020; Gössling et al., 2020; Hockings et al., 2020).

The worst-case scenario envisages a return to *tourism as usual—the "old normal."* The causal link between the pro-growth norm, top-down governance approaches, and managerial tools will continue to foster the marketization and corporatization of tourism, serving the interests and the agenda of powerful tourism industry players and governments, in the face of human, social, and environmental challenges. This return to the status quo ante, in all likelihood, will be aided by the expected scientific breakthroughs in the treatment of the disease and by the likely development and global use of an effective vaccine, affecting positively, in terms of economic growth, both the tourism supply and the demand side. The need to overcome the global economic recession strengthens big industry players, who are capable of driving the economic recovery more quickly. This development would go hand-in-hand with an increase in menial work and exploitative working conditions, undermining workers' protection and safety, fair contracts, and remuneration, as well as workplace quality. Without a doubt, the factors impacting work conditions will affect service quality. The eventual reduction of uncertainty, health risk, and government-imposed restrictions will help people regain the trust and confidence needed to travel, while the uneven distribution of revenue losses will create even greater disparity between people who maintained economic well-being and disposable income and those who struggle to find work. These processes, combined with underfunding of existing initiatives, are likely to widen the environmental and social divide. A return to "tourism as usual" is a return to tourism as a function of capitalist expansion (Fletcher, 2011). Its "logic thrives on unjustifiable value capture in situations where the externalities are absorbed by the environment, the precarious labor force, host communities, and mom and pop businesses" (Ioannides & Gyimóthy, 2020, p. 6). However, the advisability of a simple return to the "old normal" is increasingly being questioned. The tourism sector worldwide is taking such a hit that many companies will disappear, consumers will change travel habits, and destinations and their governments will have to explore alternatives. However, rectifying an uneven global system is not easy, because of path-dependent processes driven by the logic of capitalism. A path that leads to transformation in tourism can be realized if sufficient institutional innovation occurs on both the demand and supply side of tourism (Brouder, 2020).

The cautious scenario envisages a return to *tourism as un-usual—the "new normal."* The growth-led, corporate-managed, resource-intensive models of tourism development will continue under circumstances of a deeper global economic depression, such as if the global pandemic lasts longer with ongoing contagion, or becomes deadlier than originally thought. As international tourism continues to be dramatically reduced, impacting employment and income, most governments adopt massive stimulus packages to restart economies, but with a strong focus on job creation. Environmental conservation spending and regulation are reduced and weakened. Tensions are expected in rural communities because of the return of people having lost their jobs in urban areas or as a result of the increased number of people leaving the city to "telework" from the countryside. Rural areas will experience more pressure on natural resources and potential conflicts with local residents because of an increase of visitors "meeting locals and living like a local" (Richards, 2014). In the new normal, "staycation" practices (Molz, 2009) encourage people to be tourists in their home region, supporting the local economy with alternative tourism practices. This represents a significant change for people who for years have favored "fast tourism," traveling at a frenetic pace to visit cities, parks, and faraway countries. The weekend getaway with a low-cost airline flight is replaced by slower forms of tourism, where people will (re)discover the natural environments and cultural attractions their region has to offer. In a COVID-19 period that may last several years, travel solutions that propose escapes or microadventure opportunities away from crowds will be favored. The virtues of slow nature travel may be re-discovered as a means toward happiness and managing mental health (Buckley, 2020). Local businesses and communities must adapt in order to address these new market opportunities and expectations.

The more optimistic scenario envisages an unprecedented opportunity for a reboot of tourism (Niewiadomski, 2020)—we call this transition *humanistic tourism* (Della Lucia & Giudici, 2021)—that is, tourism with a strong concern for human dignity, values, welfare, and nature. While we already know what is being "destroyed," there is still much wishful multidisciplinary thinking, which overlaps significantly, about what the new tourism configuration would be.

Humanistic tourism must be strongly aligned with the tenets of *sustainability* and *humanistic management*, in order to create shared value, while dignifying people and communities (Della Lucia & Giudici, 2021). This entails being less exploitative and greedy; more environmentally sustainable; more respectful to communities and their cultures and traditions; and more mindful, fair, and compassionate (Higgins-Desbiolles, 2020a). In this vein, *humanistic tourism* envisions human flourishing as a deep shift in values (Cheer, 2020), toward revaluing and reconceiving the way we inhabit the planet and interrelate with humans and non-humans.

Humanistic tourism must be in line with the principles of *regenerative economics* (Ateljevic, 2020; Cave & Dredge, 2020): being in a right relationship with both the culture and the ecosphere in which the human economy is embedded; viewing wealth holistically in terms of well-being of the "whole" instead of mere money; being creative, innovative, adaptive, and responsive; empowering participation of individuals and groups; honoring and nurturing healthy and resilient communities and places; cultivating diversity as a source of creativity; shaping circulatory, and value-enhancing flows (of information, production, etc.); seeking balance between dichotomies.

Humanistic tourism responds to the climate crisis and endorses the shift to a carbon-neutral economy by "flattening the (growth) curve" (Prideaux, Thompson & Pabel, 2020). This approach embeds and supports the incipient *circular economy* model by adopting old and new strategies fostered by technology-driven innovations. This entails reducing, reusing, and recycling, to achieve harmony with the fragility of resources required to sustain human life; renting rather than owning, moving from in-built obsolescence to repair and reuse for other purposes.

Humanistic tourism is a return to tourism as a *social force* (Higgins-Desbiolles, 2006, 2020b): being a connector of people; bridging local communities and tourists; empowering and building greater well-being of local communities; and fulfilling wider social promises, including justice, equity, and autonomy. In this vein, *humanistic* tourism adopts a "Buen Vivir approach" (Everingham & Chassagne, 2020), requiring tourism to be small-scale, slow, and local and benefiting host communities as well as tourists to increase the well-being for all.

4. Envisioning Recommendations for Shared Value Creation

The multidisciplinary thinking about a new tourism configuration converges around the fact that the pandemic results in significant changes in how humanity perceives our planet, our relationship to nature and with one another, and with technologies. At the core of these changes, there is the notion of *conviviality* and the need of a convivial revolution (Butcher, 2020) in the economy, society, and the environment. "Conviviality" etymologically comes from the Latin word *convivium* (banquet) and, in turn, from the verb *convivere* (living together). Living together almost inevitably involves sharing meals but becomes a symbol of people gathered for common and high purposes, including the re-building of the ways in which society and the economy work. According to Illich (1973), "conviviality" designates the individual freedom realized in personal interdependence. As such, it forms an intrinsic ethical value for the reconstruction of society, so that autonomous individuals and groups may contribute to the total effectiveness of a new system of production designed to satisfy human needs, which it also determines. In other words, a convivial society and the convivial structure of tools are the basis to promote a convivial and pluralist mode of production, which ensure distributive and participatory justice.

Under this premise, the impact of the pandemic on tourism envisions recommendations to build *humanistic tourism*, based on shared value processes at the interplay between sustainability and humanistic management. These lessons revolve around clusters of interactions: *human vs. human, human vs. nature, human vs. technology, and human vs. the economy.*

Human vs. human: The human-to-human relationship is at the core of conviviality and, in turn, of tourism as a social force connecting people, facilitating sociability and associational life. In the end, it is people who travel and people who host travelers. Health concerns, home isolation, quarantine, social distancing and other measures have prevented us from meeting, face-to-face—when, how, and where we want—forcing us to distance ourselves from the people we physically encounter and to meet more often online for business as well as for social and leisure purposes. These circumstances have forced us to consider what it means to be human and what is truly meaningful and valuable for us. Health and biosecurity, sociality, trust, confidence, and self-fulfillment/fulfillment of human potential, among others, all relate to possibilities inherent in sociability, that is, cultivating and celebrating convivial life. This sense of security, belonging, community, and citizenship should be embedded in *humanistic tourism*, starting from (re)creating trust and the confidence to travel. At the core, there is a new emphasis on meaningful experiences (Smith & Diekmann, 2017): giving more value and more time (slow tourism) to self-help and self-development (well-being tourism), to human interactions within the traveling group (friends and family), and with the host community (indigenous/community-based tourism). From a business perspective, tourism operators should be more involved in facilitating social encounters and in benefiting the many people who are at the receiving end of tourism (both from the supply and demand side). Community support and well-being should become core values and objectives of the tourism sector.

Human vs. nature: The human-to-nature relationship is another dimension that pandemic-related circumstances have forced us to give increasing value and importance to, raising awareness and responsibility at both the macro-meso level and the micro one. It is now well accepted that humans' impact on the planet is destructive, leading to pollution, natural resource depletion, loss of biodiversity, and climate change. We do expect that the generation of young people who led the 2019 climate change demonstrations in support of Greta Thunberg will implement the changes that are needed to consider our relationship with nature differently. Tourism has certainly played a major role in the significant and irremediable environmental damage that has occurred over the past 30 years. The improvement of natural ecosystems and wildlife, resulting from the stark decline in international air travel, cruise travel, and overtourism caused by the pandemic, is a clear sign of the tourism vs. nature interaction. The tourism sector must work with governments and lead the charge in setting goals for decarbonization and depollution (i.e. environmental sustainability) but also by reshaping conservation programming. Convivial conservation (Fletcher & Büscher, 2020) does refer to an approach that allows humans and non-humans to live side by side in meaningful coexistence and supports and subsidizes the livelihoods of people living intimately with wildlife. The move toward slow tourism, ecotourism, and staycation practices mirrors this search for meaningful coexistence and engagement with nature. These tourism practices encourage greater environmental stewardship and sustainable and greener behavior, along with having extensive benefits for well-being.

Human vs. technology: Because of communication and marketing technology–driven democratization, the changing relationships among humans fostered by the pandemic come with an acceleration of our relationship with technology. This interaction has both bright and dark sides. Containment measures, safety, anxiety and fear issues, as well as recovery strategies, have been balanced and mitigated by the technology mediation in personal relationships and the increase of technology adoption in business of any kind. During the lockdown, nearly everyone has spent more time—if not almost all of the time—online, from work and learning to fitness, leisure, and socializing. By contrast, in the next phase of deconfinement, the desire to spend time offline may intensify with the search for digital detox retreats. On the other hand, businesses have increased the digitalization of their processes and services, from robots to artificial intelligence (AI) and drones, all in an effort to limit human contacts and to increase efficiency. For example, in the tourism sector, robot technology has been adopted to clean hotel rooms, to disinfect public spaces in airports, or to check travelers' temperature before boarding planes. This shift affects human-to-human encounters at the base of tourism as a social force and triggers substitution mechanisms. Seyitoğlu & Ivanov (2020, p.1) suggested that "robots create a technological shield between tourists and employees that increases the physical and emotional distance between them". While recognizing that COVID-19 has fostered a broader technology and robotics adoption, Zeng, Chen & Lew (2020) also argued that we increasingly rely, for business and pleasure, on video technologies that can be seen as a threat to the travel and tourism sector. Technologies overcome space distance, time, and other access constraints, providing (virtual) spaces and tools to experience conviviality and learning and sharing processes (Havitz et al., 2020). The progress made in virtual reality (VR) may help some of these technologies become marketing tools but also substitutes, for certain markets, to actual travel (Griffin et al., 2017; Griffin & Muldoon, 2020). Technology is not an end in itself. It needs to be at the service of humans, be they consumers, service providers, or community stakeholders.

Human vs. the economy: Humans make and contribute to the economy, it being based on capitalism or on other foundations. Tourism, as a function of capitalist expansion, has shown its vulnerability amid climate change, pandemics, and other challenges. The current economic/tourism model is unsustainable and needs to be rethought. The growing emergence of certified B Corporations (https://bcorporation.net/about-b-corps) demonstrates that a new generation of business leaders are paving the way toward new business practices that respect the environment and all stakeholders. A commitment to exercise economic practices that are beneficial to multiple stakeholders, including local communities and the natural environment, is essential. Strictly related, the basic tenet of economic growth must also be rethought if we want to accomplish the goals expressed in the Paris agreement to limit fossil fuel consumption. Tourism should take the lead in reinterpreting our understanding of the relationships between hosts and guests (Smith, 1989) and in managing constrained resources for the benefit of its stakeholders. These avenues for change may include cooperation, co-evolution, nested interactions, the ethics of sharing, altruism, and progress, among others.

5. Open Epilogue

The poetic fable "The Tin Forest" encourages us to act in manners favoring the emergence of a different world:

> In the middle of a windswept wasteland full of discarded scrap metal lives a sad and lonely old man. In spite of his gloomy surroundings,

> he dreams every night of a lively forest full of trees, birds, and animals. When he finds a broken light fixture that looks like a flower, his imagination is sparked. He begins to build a tin forest, branch by branch, creature by creature. In time, real birds arrive, bearing seeds, and soon the artificial forest is taken over by living vines and animals until it looks just like the forest of the old man's dreams.
>
> (Ward & Anderson, 2013)

We should not aspire to get back to what we think was "normal." The pandemic is giving us a chance to change what "normal" is and what it should be. We, tourism academics, together with industry partners and governments, can design a path toward a new era—a new era where, in the words of Intrepid Travel co-founder Geoff Manchester (2020), "the rules can be rewritten to benefit the most vulnerable people, to protect the natural world and all the animals and humans who inhabit it, and to heal our planet from the damage we've caused."

References

Ateljevic, I. (2020). Transforming the (tourism) world for good and (re)generating the potential 'new normal'. *Tourism Geographies*, *22*(3), 467–475. https://doi.org/10.1080/14616688.2020.1759134

ATLAS Tourism and Leisure Review. (2020). Tourism and the corona crisis: Some ATLAS reflections. *ATLAS, 2*, 1–95.

Bianchi, R. V. (2020). COVID-19 and the potential for a radical transformation of tourism? *ATLAS Tourism and Leisure Review*, 2, 80–86.

Brouder, P. (2020). Reset redux: Possible evolutionary pathways towards the transformation of tourism in a COVID-19 world. *Tourism Geographies*, *22*(3), 484–490. https://doi.org/10.1080/14616688.2020.1760928

Buckley, R. (2020). Nature tourism and mental health: Parks, happiness, and causation. *Journal of Sustainable Tourism*, *28*(9), 1409–1424. https://doi.org/10.1080/09669582.2020.1742725

Butcher, J. (2020). Let the good times roll (as soon as possible): Why we need a post COVID convivial revolution. *ATLAS*, *2*, 26–36.

Cave, J., & Dredge, D. (2020). Regenerative tourism needs diverse economic practices. *Tourism Geographies*, *22*(3), 503–513. https://doi.org/10.1080/14616688.2020.1768434

Cheer, J. M. (2020). Human flourishing, tourism transformation and COVID-19: A conceptual touchstone. *Tourism Geographies*, *22*(3), 514–524. https://doi.org/10.1080/14616688.2020.1765016

Della Lucia, M. (2013). Economic performance measurement systems for event planning and investment decision making. *Tourism Management*, *34*, 91–100. https://doi.org/10.1016/j.tourman.2012.03.016

Della Lucia, M., & Giudici, E. (2021). *Humanistic tourism: Values, norms and dignity*. New York: Routledge.

Della Lucia, M., Giudici, E., & Secchi, D. (2021). Reshaping tourism: Advances and open issues. In M. Della Lucia & E. Giudici (Eds.), *Humanistic tourism: Values, norms and dignity* (pp. 231–244). New York: Routledge.

Dodds, R., & Butler, R. W. (Eds.). (2019). *Overtourism: Issues, realities and solutions*. Berlin and Boston: De Gruyter.

Ellis, G., & Witt, P. (1984). The measurement of perceived freedom in leisure. *Journal of Leisure Research*, *16*(2), 110–123. https://doi.org/10.1080/00222216.1984.11969579

Everingham, P., & Chassagne, N. (2020). Post COVID-19 ecological and social reset: Moving away from capitalist growth models towards tourism as Buen Vivir. *Tourism Geographies*, *22*(3), 555–566. https://doi.org/10.1080/14616688.2020.1762119

Fletcher, R. (2011). Sustaining tourism, sustaining capitalism? The tourism industry's role in global capitalist expansion. *Tourism Geographies*, *13*(3), 443–461. https://doi.org/10.1080/14616688.2011.570372

Fletcher, R., & Büscher, B. (2020). Conservation basic income: A non-market mechanism to support convivial conservation. *Biological Conservation*, *244*(108520). https://doi.org/10.1016/j.biocon.2020.108520

Fletcher, R., Murray Mas, I., Blázquez-Salom, M., & Blanco-Romero, A. (2020). *Tourism degrowth, and the COVID-19 crisis*. Retrieved from https://politicalecologynetwork.org/2020/03/24/tourism-degrowth-and-the-covid-19-crisis/

Freeman, J. (2020). South Dakota vs. Coronavirus: Statewide shutdown is not the only way to protect public health. *Wall Street Journal*. Retrieved from www.wsj.com/articles/south-dakotavs-coronavirus-11586967498

Gardiner, B. (2020, June 18). Why COVID-19 will end up harming the environment. *National Geographic*. Retrieved from www.nationalgeographic.com/science/2020/06/why-covid-19-will-end-up-harming-the-environment/

Geneva Environment Network. (2020). *COVID-19 and the environment*. Retrieved from www.genevaenvironmentnetwork.org/resources/updates/updates-on-covid-19-and-the-environment/

Gössling, S., Scott, D., & Hall, C. M. (2020). Pandemics, tourism and global change: A rapid assessment of COVID-19. *Journal of Sustainable Tourism*, *29*(1), 1–20. https://doi.org/10.1080/09669582.2020.1758708

Griffin, T., Giberson, J., Lee, S. H., Guttentag, D., Kandaurova, M., Sergueeva, K., & Dimanche, F. (2017, June). *Virtual reality and implications for destination marketing*. Paper presented at the Travel and Tourism Research Association International Conference, Québec, Canada. Retrieved from https://scholarworks.umass.edu/ttra/2017/Academic_Papers_Oral/29/

Griffin, T., & Muldoon, M. (2020). Exploring virtual reality experiences of slum tourism. *Tourism Geographies*, 1–20. https://doi.org/10.1080/14616688.2020.1713881

Harari, Y. N. (2020, March 19). The world after coronavirus. *The Financial Times*. Retrieved from www.ft.com/content/19d90308-6858-11ea-a3c9-1fe6fedcca75

Hart, P., Heyse, L., & Boin, A. (2001). New trends in crisis management practice and crisis management research: Setting the agenda. *Journal of Contingencies and Crisis Management*, *9*(4), 181–199. https://doi.org/10.1111/1468-5973.00168

Havitz, M., Pritchard, M. P., & Dimanche, F. (2020). Leisure matters: Cross continent conversations in a time of crisis. *Leisure Sciences*. https://doi.org/10.1080/01490400.2020.1774451

Higgins-Desbiolles, F. (2006). More than an "Industry": The forgotten power of tourism as a social force. *Tourism Management*, *27*(6), 1192–1208. doi:10.1016/j.tourman.2005.05.020

Higgins-Desbiolles, F. (2020a). Socialising tourism for social and ecological justice after COVID-19. *Tourism Geographies*, *22*(3), 610–623. https://doi.org/10.1080/14616688.2020.1757748

Higgins-Desbiolles, F. (2020b). COVID-19 and tourism: Reclaiming tourism as a social force? *ATLAS Tourism and Leisure Review*, *2*, 65–71.

Hockings, M., Dudley, N., Elliott, W., Napolitano Ferreira, M., MacKinnon, K., Pasha, MKS . . . Yang, A. (2020, May). Editorial essay: COVID-19 and protected and conserved areas. *PARKS*, *26*(1), 7–24.

Illich, I. (1973). *Tools for conviviality*. New York: Harper & Row.

Ioannides, D., & Gyimóthy, S. (2020). The COVID-19 crisis as an opportunity for escaping the unsustainable global tourism path. *Tourism Geographies*, *22*(3), 624–632. https://doi.org/10.1080/14616688.2020.1763445

Manchester, G. (2020). *Travel can only rebound stronger if it rebuilds more responsibly*. Retrieved from https://bthechange.com/travel-can-only-rebound-stronger-if-it-rebuilds-more-responsibly-a76d1902b685

Mascia, M. B., & Pailler, S. (2011). Protected area downgrading, downsizing, and degazettement (PADDD) and its conservation implications. *Conservation Letters*, *4*, 9–20. https://doi.org/10.1111/j.1755-263X.2010.00147.x

Mikušová, M., & Horváthová, P. (2019). Prepared for a crisis? Basic elements of crisis management in an organisation. *Economic Research-Ekonomska Istraživanja*, *32*(1), 1844–1868. https://doi.org/10.1080/1331677X.2019.1640625

Molz, J. G. (2009). Representing pace in tourism mobilities: Staycations, slow travel and the amazing race. *Journal of Tourism and Cultural Change*, *7*(4), 270–286. https://doi.org/10.1080/14766820903464242

Niewiadomski, P. (2020). COVID-19: From temporary de-globalisation to a rediscovery of tourism? *Tourism Geographies*, *22*(3), 651–656. https://doi.org/10.1080/14616688.2020.1757749

Novelli, M. (2020). Travel at the time of COVID 19—get ready for it! *ATLAS Tourism and Leisure Review*, *2*, 13–18.

Perrow, C. (1999). *Normal accidents. Living with high-risk technologies*. Princeton, NJ: Princeton University.

Popovich, N., Albeck-Ripka, L., & Pierre-Louis, K. (2020, May 20). The trump administration is reversing 100 environmental rules. Here's the full list. *The New York Times*. Retrieved from www.nytimes.com/interactive/2020/climate/trump-environment-rollbacks.html

Porter, M., & Kramer, M. (2012). Shared value: The bridge between corporate social responsibility and corporate strategy. In A. Schneider & R. Schmidpeter (Eds.), *Corporate social responsibility* (pp. 137–153). Berlin and Heidelberg: Springer.

Prideaux, B., Thompson, M., & Pabel, A. (2020). Lessons from COVID-19 can prepare global tourism for the economic transformation needed to combat climate change. *Tourism Geographies*, *22*(3), 667–678. https://doi.org/10.1080/14616688.2020.1762117.

Pritchard, M., Havitz, M., & Howard, D. (1999). Analyzing the commitment-loyalty link in service contexts. *Journal of the Academy of Marketing Science*, *27*(3), 333–348. https://doi.org/10.1177/0092070399273004

Reason, J. (1990). *Human error*. Cambridge: Cambridge University.

Richards, G. (2014). Creativity and tourism in the city. *Current Issue in Tourism*, *17*(2), 119–144. https://doi.org/10.1080/13683500.2013.783794

Seraphin, H., Sheeran, P., & Pilato, M. (2018). Over-tourism and the fall of Venice as a destination. *Journal of Destination Marketing & Management*, *9*, 374–376. https://doi.org/10.1016/j.jdmm.2018.01.011
Seyitoğlu, F., & Ivanov, S. (2020). Service robots as a tool for physical distancing in tourism. *Current Issues in Tourism*, 1–4. https://doi.org/10.1080/13683500.2020.1774518
Smith, M. K., & Diekmann, A. (2017). Tourism and wellbeing. *Annals of Tourism Research*, *66*, 1–13. https://doi.org/10.1016/j.annals.2017.05.006
Smith, V. (1989). *Hosts and guests: The anthropology of tourism*. Philadelphia: University of Pennsylvania.
Venette, S. (2003). *Risk communication in a high reliability organization: APHIS PPQ's inclusion of risk in decision making*. Ann Arbor, MI: UMI ProQuest Information and Learning.
Ward, H., & Anderson, W. (2013). *The tin forest*. London: Templar Publishing.
Weaver, D. B. (2000). A broad context model of destination development scenarios. *Tourism Management*, *21*, 217–224.
Weaver, D. B. (2012). Organic, incremental and induced paths to sustainable mass tourism convergence. *Tourism Management*, *33*(5), 1030–1037.
WTTC. (2020). *Recovery scenarios 2020 & economic impact from COVID-19*. Retrieved from https://wttc.org/Research/Economic-Impact/Recovery-Scenarios-2020-Economic-Impact-from-COVID-19
Zeng, Z., Chen, P. J., & Lew, A. A. (2020). From high-touch to high-tech: COVID-19 drives robotics adoption. *Tourism Geographies*, 22(3), 724–734. https://doi.org/10.1080/14616688.2020.1762118

Index